PERSONAL CONTENT EXPERIENCE

PERSONAL CONTENT EXPERIENCE

Managing Digital Life in the Mobile Age

Juha Lehikoinen

Antti Aaltonen

Pertti Huuskonen

Ilkka Salminen

John Wiley & Sons, Ltd

Other Wiley Editorial Offices

John Wiley & Sons Inc., 111 River Street, Hoboken, NJ 07030, USA

Jossey-Bass, 989 Market Street, San Francisco, CA 94103-1741, USA

Wiley-VCH Verlag GmbH, Boschstr. 12, D-69469 Weinheim, Germany

John Wiley & Sons Australia Ltd, 42 McDougall Street, Milton, Queensland 4064, Australia

John Wiley & Sons (Asia) Pte Ltd, 2 Clementi Loop #02-01, Jin Xing Distripark, Singapore 129809

John Wiley & Sons Canada Ltd, 6045 Freemont Blvd, Mississauga, ONT, L5R 4J3, Canada

Wiley also publishes its books in a variety of electronic formats. Some content that appears in print may not be available in electronic books.

Anniversary Logo Design: Richard J. Pacifico

British Library Cataloguing in Publication Data
A catalogue record for this book is available from the British Library
ISBN 978-0-470-03464-4 (PB)

Typeset by SNP Best-set Typesetter Ltd., Hong Kong

This book is printed on acid-free paper responsibly manufactured from sustainable forestry in which at least two trees are planted for each one used for paper production.

Contents

Foreword

This book is an innovative study of user experience design and analysis for the age of mobile computing and communication. Phones, PDAs, media players, and other devices are ubiquitous platforms for relationship-building, learning, entertainment, commerce, and socializing. The first generation of mobile studies focused on traditional aspects of user-interface design, human factors, and usability. The coming generation focuses on usefulness and appeal. Content and controls are combined in complex and novel ways. New metaphors, mental models, and navigation combine with changing techniques of interaction and appearance to produce new multi-modal, multi-media, and multi-user forms and formats that concentrate as much on the content as on the controls. Today, the amount of mobile personal content is exploding and is tightly coupled with increasing mobility, personal memories, and daily communications with others. Consumers, as well as professionals in academia and industry, are just becoming aware of the consequences.

This book sets the stage for understanding personal content experience and uncovers the consequent challenges. The authors take a multidisciplinary approach ranging from user-centered design to mobile software development, from information management to thorough analysis of media, context, and metadata from the consumer point of view. The book is among the first of its kind to consider the full range of issues, from software architecture to end-user needs for mobile personal content.

The authors introduce and explore issues such as context capture, user interfaces for continuous mobile use, user-interface design for mobile media applications, complex metadata, virtual communities, and ontologies. The also discuss user interactions and behavioral patterns with personal content, resulting in a lifecycle model for analyzing media devices, services, applications, and user interfaces. Finally, they examine personal content from a number of viewpoints and present a mobile prototype software framework for personal content management, complete with extensible ontologies, that realizes the requirements of an enjoyable personal-content experience. Consequently, the book contains both theory, in the form of frameworks and models, and practice, such as hands-on examples, application concepts, and software architecture descriptions.

The authors, each with several years of professional experience in mobile personal content, provide clear insight into this new, emerging topic, providing both industrial insight and scholarly analysis. Because the book contains a detailed description of proposed, extensible software architecture targeted at content management in mobile devices, these essential details will benefit anyone developing mobile content-intensive applications and services. The authors provide clear and practical guidance to this new field of research and development, making the book applicable to a wide audience, including practitioners in mobile technology and digital content management, media-intensive application developers, content creators and distributors, academic researchers, and faculty in computer science and multimedia.

We are entering a new phase in the development of mobile content, applications, and services. This book helps us to see the patterns of this evolution.

Aaron Marcus, President,
Aaron Marcus and Associates, Inc.,
Berkeley, California, USA

Acknowledgements

The authors wish to express their gratitude to the following individuals for their valuable contributions towards this book: Juha Arrasvuori, Sam Bucolo, Gunner Danneels, Jim Edwards, Anna Hakala, Jussi Holopainen, Jyri Huopaniemi, Quinn Jacobson, Matti Karlsson, Imre Kiss, Hannu Korhonen, Sanna Korpi, Timo Koskinen, Markku Laitkorpi, Ora Lassila, Kari Laurila, Jaakko Lehikoinen, Aaron Marcus, Jouka Mattila, Joe McCarthy, Hannu Nieminen, Marja Niska, Tero Ojanperä, Dana Pavel, David Racz, Erika Reponen, Mika Röykkee, Larry Rudolph, Risto Sarvas, Mirjana Spasojevic, David G. Stork, Toni Strandell, and Kai Willner.

Special thanks must go to Tero Hakala for providing and editing much of the photographic material in this book, to Juha Kaario for jointly developing the GEMS model, and to Jussi Impiö for vividly portraying the personas.

In addition, the authors would like to thank their families for their support: Tiina, Anniina and Eemeli; Jonna, Anni and Chili (woof); Marja, Petteri, and Valtteri; Annu and Aamu.

We wish to thank the staff of our publisher for smooth collaboration: Birgit Gruber, Sarah Hinton, Richard Davies, and Rowan January at John Wiley & Sons.

No animals were harmed during the making of this book, but our laptops did receive much punishment. Rest in pieces.

Juha Lehikoinen
Antti Aaltonen
Pertti Huuskonen
Ilkka Salminen

Tampere, Finland

List of Abbreviations

2D	two-dimensional
3D	three-dimensional
3G	third-generation mobile phone technology
3GPP	Third-Generation Partnership Project
AAC	Advanced Audio Codec
Adobe XMP	Adobe Extensible Metadata Platform
AI	artificial intelligence
ATM	automated teller machine
ATRAC	Adaptive Transform Acoustic Coding
API	application programming interface
ASCII	American Standard Code for Information Interchange
AVC	Advanced Video Coding
bit	binary digit
CBR	content based retrieval
CCC	consumer created content
CD	compact disc
CE	consumer electronics
CE	context engine
CEP	Context Exchange Protocol
CLM	content lifecycle management
CM	communications manager (in context engine)
CMS	content management system
CPU	central processing unit
DBMS	database management system
DCF	design rule for camera file system
DCMI	Dublin Core Metadata Initiative
DLL	dynamic link library
DLNA	Digital Living Network Alliance
DRM	digital rights management
DV	digital video
DVB	digital video broadcasting
DVB-H	digital video broadcasting – handheld
EFF	Electronic Frontier Foundation
EP	extended play record
EPG	electronic program guide
EPOC	excess post-exercise oxygen consumption
EXIF	Exchangeable Image File Format

GB	gigabyte
GEMS	get enjoy maintain share
GPRS	General Packet Radio Service
GPS	Global Positioning Service
GSM	Global System for Mobile communications
GUI	graphical user interface
GUID	globally unique identifier
H.264	a video compression standard, part of MPEG-4 Part 10
HCI	human–computer interaction
HSUPA	High-Speed Uplink Packet Access
HTML	Hypertext Markup Language
HTTP	Hypertext Transfer Protocol
Hz	hertz
IC	integrated circuit
ID3	tagging standard for MP3 files
IETF	Internet Engineering Task Force
IIM	Information Interchange Model
IM	instant messaging
IMT	Internet Media Type
IP	Internet Protocol
IPTC	International Press Telecommunications Council
IR	information retrieval
IRC	Internet Relay chat
ISO	International Standards Organization
ITU	International Telecommunication Union
ITU-T	ITU Telecommunication Standardization Sector
JEITA	Japan Electronics and Information Technology Industries Association
JPG, JPEG	Joint Photographic Experts Group
LAN	local area network
LCD	liquid crystal display
LED	light emitting diode
LP	long play record
MARC	machine-readable cataloguing
Mbit	megabit
MdE	metadata engine
MESH	medical subject headings
MIDI	Musical Instrument Digital Interface
MMOG	massively multiplayer online game
MMORPG	massively multiplayer online role playing game
MMS	multimedia messaging service
MP3	MPEG-1 audio layer 3, an audio compression standard
MPEG	Moving Picture Experts Group

MPEG-1, MPEG-2, MPEG-4	audio and video compression formats developed by MPEG
MPEG-7	metadata format designed by MPEG
ms	millisecond
MUD	multi-user dungeon
MUPE	multi-user publishing environment
NFC	near field communication
NTP	Network Time Protocol
OCR	optical character recognition
ODP	open directory project
OMA	Open Mobile Alliance
OS	operating system
P2P	peer-to-peer
PC	personal computer
PCE	personal content experience
PCM	personal content management
PDA	personal digital assistant
POI	point of interest
PIM	personal information manager
PSP	playstation portable
PVR	personal video recorder
QWERTY	acronym describing layout of standard typewriter keyboard
QVGA	quarter VGA
RAM	random access memory
RDF	resource description framework
RFC	request for comments
RFID	radio frequency identification device
RPG	role playing game
RSS	Really Simple Syndication
RSVP	rapid serial visual presentation
RTF	Rich Text Format
SD	Secure Digital (memory card)
SDK	software development kit
SIM	subscriber identity module
SMS	Short Message Service
SNTP	Simple Network Time Protocol
SQL	Structured Query Language
SUI	speech user interface
SVG	scalable vector graphics
SyncML	Synchronization Markup Language
TFT	thin film transistor
TIFF	Tagged Image File Format

TTS	text to speech
UCD	user centric design
UI	user interface
UPnP	universal plug and play
URI	uniform resource identifier
URL	uniform resource locator
USB	Universal Serial Bus
USD	United States dollars
UWB	ultra-wide band
VGA	Video Graphics Array
VOIP	Voice-Over-Internet Protocol
VPN	virtual private network
VR	virtual reality
W3C	World Wide Web Consortium
WAV	Waveform audio format
Wi-Fi	a WLAN that conforms to IEEE 802.11b standards
WIMP	Windows, Icons, Menus, and Pointing Devices
WLAN	wireless local area network
WMA	Windows Media Audio
WWW	world wide web
XML	Extensible Markup Language
xDSL	X digital subscriber line
XGA	Extended Graphics Array
XviD	implementation of a MPEG-4 video codec
ZUI	zoomable user interface

Prologue

Personas, imaginary people that model and present the essential user segments, are a popular tool in user-centred design. They are formed on the basis of studies and observations of users, for aiding designers and guiding the creation of features, interaction, UI navigation, and terminology in products.

In this book we use personas for illustrating the concepts, trends, and technologies which are discussed. Throughout the text, we present brief moments in the personas' lives and experiences related to personal content. The moments highlight the terms and concepts presented in the text.

Let us introduce these personas:

ANGIE, 16, STUDENT

Angie is a student at the local Belleville High, near London. She lives in a nice house with her parents and her younger brother Adam, 14. Her uncle Ron is also close, but physically far away living in Nevada. They keep in touch by e-mail and occasional video chat sessions.

Angie has little money to spend, but still enjoys an unusually good selection of electronics, as her dad receives good staff discounts at Dixons. Her otherwise girly room is stocked with a laptop, music players, a video camera, a digicam, and a mobile phone or two.

She has a relatively active social life, often hanging out in cafés with the girls. Her boyfriend Eddie goes to the same school. They often IM in the wee hours, sometimes seeing over video. She records all the sessions, to Eddie's dismay.

Angie keeps a private diary, posts pictures on the Net, IMs, e-mails, and text messages with her friends. She downloads music occasionally, but gets most of her MP3s from her boyfriend, who is into P2P file sharing.

BOB, 28, ACTIVIST

Bob leads a double life. In the office he plays the role of a web designer, dutifully crafting page content and the occasional script for nondescript companies.

After office hours Bob turns to his passion, the AARGH.org movement (Activism Against Rotten Globalization). He is part of a mildly radical activist cell which is tightly connected virtually, although the members live all over the globe. The majority of his closest allies come from the UK and France. They usually gather physically together only for protests, which may take place once or twice a year. The rest of the time they meet over the Internet, using all the channels available: e-mail, IM, VOIP, video calling, wiki pages, shared storage, and whatnot. When Bob and his allies gather for a demonstration they apply all of the available communication techniques.

Besides activism, Bob leads a quiet life, does not drink or smoke, listens to classical music, and has no pets. The only thing that worries his good landlady is the health of her dear tenant: "Does the poor lad have any friends at all, spending his time with all those computer screens and mobile phones in his room?"

CATHY, 40, ARTIST

Cathy is a self-employed graphics artist. Ever since she learned to hold a crayon, art has been her passion. Over the years she has explored various mediums and methods of expression, but these days she is back to her beloved topic, watercolour paintings of flowers and herbs.

Cathy taught herself to use computers at her local Open University, with the help of the old flame of hers, Steve. Now she is able to scan her paintings and send them via e-mail to the Web shop. Her work is displayed at a gallery on Crow Street, London, where she shares space with a dozen other artists. They have established the Web shop where the artwork is on sale. The sales are slow, but the rent is cheap, thanks to the rich old lady that funds the gallery as a hobby.

Cathy maintains an overflowing phonebook filled with details of fellow artists, many of whom have become her close friends over the years. Too bad the phone bill is overflowing too.

DAVE, 56, PLUMBER

Dave, a professional plumber, runs a small garage company with two other guys. They are long-time professionals and have become close friends over the years. They focus on the job, trusting all accounting, invoicing, and other office activities to Janet, who runs an accountancy company. Most of their files are kept in paper form at the back of their garage.

Dave is not a gadget freak, and avoids computers whenever possible. However, he has recently switched to a more modern mobile phone that can access Internet betting sites. Obviously he has had a mobile phone for years, as much of their business involves fixing leaks in the middle of some other task. Clients' calls must be taken a.s.a.p., although these days the business is so good that he has to say "No" a little too often.

After work, Dave is in no hurry to get home and watch the tube with his nagging wife, so he often goes with the guys to the local pub for a few pints and darts. The other guys are into football, but Dave follows the equestrian scene. He frequently bets on horses and has decades of experience on evaluating the odds. In fact, that was a good source of income for his family when business was poor.

EDDIE, 16, STUDENT

Eddie's got some attitude, man. He feels right at home in the local hip-hop scene, although Belleville is not quite the happenin' place like London. Luckily there are a couple guys in the 'hood who get it as it comes to good music. Like, Anvil, Snoopy Foggy Fog, 60 Pence, you know. With the guys they can swap stuff, like MP3s and videos. Plus a couple warez ripped from P2P sites, occasionally, but nobody saw nothin'.

Lately Eddie has been spending more and more time with Angie, the blonde in the next class. They met at a party at Angie's place, and now they are sharing most of their lives (and most of their digital content).

Eddie has got deeply involved in an online game called Surreal Shards, where millions of players meet online in a shared world. Eddie & co will soon be in trouble if they cannot find some real world funds to support their virtual troops.

In addition, they play downloadable games with their mobile phones. They have access to a range of titles, mostly cracked commercial games downloaded off hacker sites on the Internet. Good thing that his dad gets him the latest phones, but the old man still has no idea how his gadgets are used.

Chapter 1: Digital Memories and the Personal Content Explosion

> application   design   device
> **digital experience**
> information   life   management
> metadata **mobile** UI **user**

This book sets the stage for understanding *personal content experience*. This is a vast topic, tightly coupled with increasing mobility, our personal memories, and daily communications with others.

We introduce and explore issues such as context capture, user interfaces for continuous mobile use, UI design for mobile media applications, metadata magic, virtual communities, and ontologies. User interactions and behavioural patterns with personal content are also covered, resulting in a "GEMS" lifecycle model for analysing media devices, services, applications, and user interfaces. We examine personal content from a number of viewpoints, and present a mobile prototype software framework, complete with extensible ontologies, which realizes the requirements of an enjoyable personal content experience.

Personal Content Experience: Managing Digital Life in the Mobile Age   J. Lehikoinen,
A. Aaltonen, P. Huuskonen and I. Salminen © 2007 John Wiley & Sons, Ltd

The reason why *you* will become attracted to this topic is evident, even if you are not a researcher or practitioner in one of the relevant fields. Personal content experience – capturing, enjoying, and sharing our digital past and present – inherently concerns us all to a large degree. Understanding this vibrant area of research, development, and everyday life is what this book is about.

1.1   Digital Us

You are becoming digital. Already, your life is digital in many ways. Increasingly everything you do with other people, and the world around you, is digital. Like it or not, you are part of a great information whirl-wind that weaves your digital life patterns with those of myriads of other people – and machines.

A remarkable part of your digital life is being stored by authorities, your employer, hospitals, the loyal customer program of your favourite chain store, and countless others. This information *about you* identifies you, stores your medical history, skills, salary history, behaviour as a consumer, and, indeed, faults and committed crimes, if any. You do not have much control over it, except for some reviewing and reclama-tion cases. This is essential information about you, and the conse-quences of possible errors may be severe. Yet you are not responsible for its maintenance.

The information described above is information about you. Yet it is not *yours*, since you do not possess it nor can freely control it. When discussing your digital life, the other side of the coin is *your* personal information, something that is important to you and that you possess. This is information that we refer to as *personal content*,[1] the topic of this book.

Your personal content is among the first aspects of your life to go digital. Your daily communications, photographs, favourite music and movies, heart rate and step count logs, contacts stored in your mobile phone, and frequently travelled routes are, among other content types, your personal content. All digital, all existing in a virtual realm consist-ing of a series of 0s and 1s, all created either implicitly or explicitly with a myriad of digital devices, such as camcorders, GPS navigators, or mobile phones. This is content that you are responsible for taking care of; it is content that is meaningful to you; it is content that you

---

[1] In many contexts the term *personal information* is associated with office activities, such as calendar, to-do items, and e-mail as implied by the common acronym PIM (Personal Information Management). This is why we prefer the term "content".

have interacted with. It is content that forms, in increasing amounts, the record of your own invaluable personal history. As your personal content is getting digital, the ways you create, share, and experience it are also about to change.[2]

The transition from analogue (or physical, if you prefer) to digital personal content is much more fundamental than it first appears. Digital content is not merely an addition to analogue personal content, but rather is the dominant way to manage and experience personal content – or most parts of it. Once the content is in digital form, it can be manipulated in countless ways, reused, distributed, and shared with practically no extra effort, and experienced through a variety of different devices. As a consequence, new behavioural patterns will appear as the way people capture, share, and re-live their experiences and personal history changes.

Consider an obvious case – photography. The digital change has become strikingly evident in this domain. The sales of digital cameras have surpassed the sales of traditional film cameras, and today the sales figures are astonishing – for instance, during the first seven months of 2006, the sales of digital cameras grew by 20.3%, while the sales of film cameras went down by 67.1%.[3] This implies that the services offering analogue film development are becoming rare and expensive, which further boosts the trend of personal content digitalization.

As a consequence, people are now taking digital photos instead of film photos. But what do they do with the photos? Some people use digital cameras just as they used film cameras. They print out the photos, and then sweep the memory card clean as if changing the film, and continue shooting. However, many people transfer the images to a PC, a set top box, or some other digital storage device. Even the photos that are not printed are in most part retained. Over the years, this collection of personal information becomes invaluable. It is content that people wish to access, share, sometimes re-organize, but definitely not lose.

But it is not just photographs that we are talking about here. While photographs often depict important bits and pieces of our personal histories, so do *messages* sent and received; *personalization settings* for our devices; self-created and maintained *characters in an online game*; *ratings* given to a certain song; *health information* measured and recorded by devices such as heart rate monitors; *music* listened to on a special occasion; and so on.

---

[2] Content created by the users themselves is also sometimes referred to as "Consumer Created Content", or CCC.

[3] http://www.imaging-resource.com/NEWS/1158601458.html

Indeed, while self-created content is an important type of personal content, it is not the only one. We assume that all content that you have interacted with and that is in your possession, is personal in nature (personal content is defined in section 3.1). The personal content experience is not only about *creating* and *storing* content, but is also about *enjoying* it.[4]

Another important aspect of personal content experience is the relationship between the pieces of personal content and general databases, because storing the information in databases makes it possible to associate personal events and interactions with other events that have taken place. This opens up new horizons for defining what a personal content experience is, especially when people wish to relive past experiences.

Content digitalization enables new ways to access memories and share them with other people. Sharing and experiencing together connects personal content tightly with communication in general. With the rapid spreading of mobile phones that enable instant sharing of experiences, we are currently witnessing one of the most remarkable transformations in human communication history and, at the same time, a fundamental change in personal content. Thus, we analyse these two emerging, heavily intertwined trends:

1. the increased mobility of people and their devices, driven by mobile communications and mobile devices for experiencing content; and

2. the explosive growth and increasing importance of personal content.

Together these two trends are likely to fundamentally and persistently change many aspects of our daily lives.

Content management is far from easy. The devices are not easy to operate, they do not connect seamlessly to each other, and information is all too easily lost, not to speak of all that navigation and browsing inside countless folders to locate a desired content object. There is indeed a need for further research and development in this area.

---

[4] Contrary to common practice, we prefer the verb *enjoy* to *consume*. Enjoying captures the essence of a wide variety of interactions with personal content, while consuming has a passive tone to it.

1.2   You and This Book

Although *content management* is a term often associated with information managers and Web content managers in enterprises, personal content must be manageable by anyone. The term "personal" provides the point of origin for all discussions in this book. This is the primary reason why we prefer to speak of personal content *experience* rather than personal content *management*.

Unfortunately, the field of personal content experience is as tangled as any. The terminology does not exist, there are no theoretical frameworks available to understand the topic in a broader context, and the term brings about twice as many opinions as there are speakers. And yet the domain is expanding in the consumer domain like a wildfire, with more and more technologies, products and services brought to the confused consumers. Therefore, the purpose of this book is to set the stage for understanding, facilitating, and enhancing personal content experience.

Several essential factors for a satisfying personal content experience can be identified. These factors are discussed in detail throughout the book, and can be summarized as follows:

- Instantaneous *ad hoc* access to one's own content with sufficient quality;

- Sharing content and vivid re-living of experiences through content;

- Trustworthiness and privacy – confidence that personal content will be kept safe and preserved;

- Joy of (re-)discovering;

- Seeing the object as a part of whole (relations);

- Exploiting mobility, metadata, and context to make content smart.

The above list is just the tip of the personal content iceberg. It is not only individual users or user groups and their experiences and behaviour that we will discuss; and it is not only the essential technologies that we will show are needed to put all the bits and pieces together. We will also address many societal issues, such as social acceptability and privacy; copyright and liability laws; issues related to mass media and media (re)production; new distribution paradigms, such as viral marketing and super-distribution; new or enhanced business opportunities, such as online services for content analysis, processing, and filtering; and so on. Because personal content is inherently tied to many aspects of our culture and society, it should not be studied separately. We will try to address the *personal content ecosystem* – not only the content itself, but also the variety of human behaviours and processes related to it, and the technical solutions that are needed to support such processes.

We cannot cover all aspects related to personal content experience in detail (it would require a series of thick books) but will focus on some key areas, and then briefly introduce some related issues.

We will assume a consumer-oriented view: when we introduce a technology, we will also show how it relates to the life of the design personas introduced in the Prologue. We strongly believe that it is possible to build systems that are understandable by the average person in terms of purpose, use, and outcomes, even though they may contain non-trivial mechanisms that handle complex data streams. The *user-centred design* approach emphasises people over technology.

This book combines views of personal content, consumer-oriented media and content management technologies, global megatrends, user interfaces, and software design. We will discuss the components of personal content management systems, ranging from system and database levels to middleware, applications and services, user interfaces, and users.

The topic of personal content spreads across several disciplines and professions, and we explore them from several different points of view. The aim is to identify the aspects that are critical for the general public and discuss the consequences they imply to the system design, application design, and user interface design. Beyond technical aspects, we also discuss the changes in use patterns and social behaviour, provoked by the mobile personal content phenomenon, and show that the impact is more fundamental than it may first appear.

The purpose of this book is twofold. On one hand, we make practical statements on how certain essential technical issues can be addressed and solved, and provide a reference software architecture design for a mobile personal content management system, coupled with comprehensive design drivers and criteria for user interfaces designed for content management in mind. These aspects are backed up with results, which we have gained from research – analysing societal, cultural, and behavioural trends in addition to technology evaluations. From this point of view, this book is best suited for software developers, user experience and interface designers, team managers, and other practitioners and professionals dealing with application and system development, especially those related to mobility, consumer media, and personal content management.

On the other hand, besides challenges in technology, user interfaces, and societal and behavioural concerns, the personal content explosion provides huge potential in new innovations in related services and business models. Thus, our intention is to increase awareness and understanding of this broad topic, and to boost innovation by challenging prevailing assumptions and discovering unexpected aspects of personal content as a research and business domain. Therefore, the book makes good reading for technology evangelists, lecturers addressing content and media, innovators, and those seeking new business opportunities within the personal content domain. We do not provide ready-made "become-a-millionaire" solutions, but rather whet the appetite with a variety of thought-provoking approaches from different points of view.

Regardless of your background, we sincerely wish that you, the reader, will become inspired, educated, disturbed, influenced, or otherwise moved when reading these pages. We also hope that this book will act as a booster for addressing this vast interdisciplinary research and business domain.

1.3   Contents at a Glance

The rest of the book is structured as follows:

Chapter 2: Trends Towards Mobility
Chapter 2 begins by introducing mobility and several aspects that tie mobile use and personal content together. We present a concise history of mobility and content digitalization, discuss several technology developments that are relevant in the context of personal content experience, characterize various mobile devices, and reveal why manufacturers are frantically trying to cram every possible function into your watch. Together, these trends paint a picture of how personal devices will evolve in the mobile years to come.

Chapter 3: Mobile Personal Content Uncovered
Chapter 3 focuses on personal content and its inherent characteristics. We define the term personal content, and provide methods for content categorization based on a number of factors. We analyse personal content from a number of viewpoints, and discuss the effects mobility has on the content. Finally, we present a high-level model for addressing different user-level aspects of personal content management. We call this model "*Get-Enjoy-Maintain-Share* (GEMS)", and show how it

can be applied in analysis and design of mobile devices, applications, and services related to personal content.

Chapter 4: Metadata Magic

In Chapter 4, we discuss metadata (data about data), one of the most important technology enablers related to personal content experience. We inspect metadata from several points of view, including existing metadata formats and frameworks, user benefits, the relationship between metadata, mobility, and context-awareness, and metadata management implementation issues. We also address closely related topics, such as increasing the semantics of content.

Chapter 5: Realizing a Metadata Framework

The beginning of Chapter 5 introduces the challenges related to distributed mobile content, and points out that metadata management is the key to solving these challenges. We discuss why metadata management cannot be left to applications alone but why a system-wide metadata framework is needed. We also discuss different types of metadata available in mobile devices. We then move deeper into metadata ontologies to ensure interoperability between applications and show how to create future-proof metadata taxonomies (and why!). We also present the design and implementation of a prototype content management framework, targeted especially at mobile use. Finally, we show how the framework functions in a mobile device.

Chapter 6: User Interfaces for Mobile Media

Chapter 6 focuses on mobile personal experience from the end-user's perspective. We start by providing a brief overview to users and then discuss how the user can interact with mobile personal content from various points of view. We discuss mobile user interface design to shed light on how user interfaces can be created based on the user's tasks. Finally, we illustrate what are the related tasks, and what is required from the UI perspective.

Chapter 7: Application Outlook

Chapter 7 combines what has been presented in the preceding chapters, and discusses application and service design for experiencing personal content. We present some general features of mobile applications and discuss several application types, and introduce existing implementations, while analysing their strengths and weaknesses. We focus on some contemporary and upcoming content-rich applications in the domains of location-based services, content sharing, games, fitness, and others. Our purpose is not to merely present the applications, but to gain an understanding of the underlying features and regularities more generic in nature.

Chapter 8: Timeshifting Life

Chapter 8 takes a look at the future. We try to identify trends and weak signals that can be extrapolated from the topics presented, and make assumptions about things to come and how to react to them. We also present some future requirements, and technological enablers that could potentially fulfil them. We also discuss potential threats and risks related to the digitalization of our personal histories. We consider whether it is desirable to be able to store almost everything in one's life versus discrete fragments, resulting in what we refer to as *timeshifting life*.

Now that we have introduced the contents, let us continue with the trends that have brought us to where we, *the new nomads*, are today.

Chapter 2: Trends Towards Mobility

We are living in a move mania. We are constantly coming or going. We fly to meetings overseas; we commute for hours; we prefer exotic faraway places for our holidays. We spend hours and hours in cars, trains, planes, hotels, airports, and other transitory places. We are becoming nomads – again.

This chapter explores us, *the new nomads*, by observing how we make use of a range of mobile devices, and put up with the compromises required by mobility. We discuss the technology advances, including miniaturization and the rise of wireless connectivity. We discuss the ways mobility influences content, using mobile music as a typical case.

Personal Content Experience: Managing Digital Life in the Mobile Age   J. Lehikoinen,
A. Aaltonen, P. Huuskonen and I. Salminen © 2007 John Wiley & Sons, Ltd

2.1   The New Nomads

Our ancestors were hunter gatherers. So are we – but not for food. We hunt after ever better, deeper, faster, richer experiences. The curiosity and appetite for something new carries us forwards under the weight of heavy luggage, and keeps us waiting patiently in immigration lines. And while we are on the way there – or already there – we want to stay connected to our people in the global village. Our global family.

It was not always like this. In the past people relied on their local communities. Earlier human societies gathered together at water holes and in caves to share necessities such as food, shelter, and tools, and later for purposes of trade and culture. Towns and cities were born out of the economy of concentrating many common functions in centralized places and also out of the human need for socializing. This trend is escalating. We now have megacities with tens of millions of inhabitants, and there is no end in sight (Mitchell 1999).

This trend towards urbanization is related to the need for sharing physical goods. As we move towards societies that are based on virtual goods (that is, information), the need for being physically close becomes less. In an information society, we can move around more freely. Many of the aspects of our lives that have previously required being in specific places can now be dealt with remotely, or while we are on the move.

Salesmen and reporters have always been on the move. They will remain so for as long as they profit from being physically present in

Figure 2-1. Urban crowds in Tokyo, a mobile megacity.

remote places or with remote customers. Other professions are inherently mobile. Some workers, such as cleaners and gardeners, must be physically present to do their work. Many other jobs are now relying on more information and less on physical items.[1] Accordingly, these jobs can be carried out remotely wherever access to information is required.

Telecommuting is becoming practical thanks to the boom in data communications. The Internet is now the *de facto* vehicle for voice and video conferencing, sharing documents, collaborative editing, or just plain chatting. Internet access has improved greatly in the past few years, particularly thanks to xDSL and cable (at the time of this writing), and mobilized by WLAN, GPRS, 3G, and other wireless data channels.

A major agent for the change is the mobile phone. It has permeated western societies rapidly, and is marching on in the developing world. In 2006, there were more than 2.6 billion mobile subscribers globally, and the number is expected to increase to 4 billion in the next few years.[2] Mobile voice access is now surpassing fixed telephony globally, as shown in Figure 2-2. Not only the western world, but also the developing populations (such as rural China, Africa, and India) witness rapid mobilization of their telephone service. For many practical purposes, we can now consider mobile telephony as the norm in urban settings.

Not only work, but also leisure is changing. Global travel, particularly for holidays, has been increasing throughout the 20th century in pace with increases in income and free time. This trend may continue until, for instance, the scarcity of natural resources drives prices of energy radically higher, or other global developments, such as biological threats, render travel risky.

2.1.1   Five-Second Attention Span

Improved communications lead to a situation where it is possible to carry out both work and leisure activities absolutely anywhere. One can read e-mail on a train; bid for an eBay[3] item in the middle of a meeting; agree to a time for golf while shopping for milk. Indeed, the new freedom of time and place seems to change our lives at a fundamental level.

---

[1] Bits instead of atoms, as Negroponte (1995) famously pointed out.
[2] http://www.pcworld.com/article/id,127820-c,researchreports/article.html
[3] http://www.ebay.com/

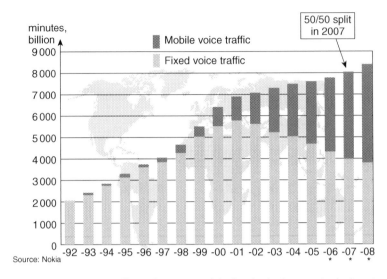

Figure 2-2. *Mobile voice traffic is the norm, while fixed telephony is in decline (repro-duced by permission of © Nokia).*

Increased mobility of people and data leads to overlapping and fragmentation of human activities. This has multiple consequences to our lives, so we need to become better at multitasking, something that humans are traditionally not very good at. It is the *protean self* we are talking about – being fluid, resilient, on the move (Lifton 1999). With computer help, we can handle more data and events within a given time, although our thought processes may not be improving. Time will tell if this will allow people do more – or just touch more topics superficially (Gleick 2000).

The average human attention span appears to be getting shorter, at least judging from the mass media. For example, TV programmes need to have enough dynamics to stop the audience getting bored, which is evident when comparing them to shows from the 1950s – the pace of the old shows seems very slow to us now. Thus today's TV requires something new every two or three seconds. The restless visual rhythm of music videos and action films has infiltrated other media. Magazine columns, Web pages, and news clips, must all be brief and to the point if the readers' interest is to be retained. Casual mobile games need to be quick to learn and easy to play. RSS feeds condense the Web into quick, brief snippets of up-to-date news. The continuous attention to mobile devices fragments to four-to-eight second bursts (Oulasvirta *et al.* 2005). Consequently, content targeted for mobile devices, such as news, videos, animations, etc., has to be designed for such short attention spans.

The notion of *gaps in life* implies that mobile content is particularly interesting while we are in between other activities. This necessarily shapes mobile content and the devices used to access it. For instance, it is not very enjoyable to view a full-length movie on a small hand-held screen (just try it!), but watching a two-minute trailer of the same movie feels fine while waiting for the bus.

2.1.2 Continuous and Nomadic Mobile Use

Mobile use can be either continuous or nomadic (Rosenberg 1998). Continuous refers to using the mobile device while actively moving, whereas nomads can be stationary for long periods of time. For instance, when a typical corporate traveller establishes her mobile office in a hotel where she is staying for a few days, she is not at her home base, but she is not altogether mobile all the time either – she is a nomad camping out temporarily. She can invest some effort in making her environment more convenient for the limited time she will spend there. She may wish to struggle through the ordeal of arranging office network access from the hotel, which she would not bother to do if she was going to be there only for an hour or two. Some of the requirements for continuous mobile use can thus be relaxed for nomadic users.

Nomadic situations also highlight people's desire for calm and serenity, even for a short while, to be able to immerse themselves in something other than the real world. For instance, after running madly through the whole day, it feels relaxing to take the MP3 player out of your pocket, sit down, put on the headset, start listening to your favourite band – and close your eyes.

Figure 2-3. The tools of a nomadic office worker.

HIGH-LEVEL RESEARCH

Bob felt exhausted after the delayed flights, the packed train, and the horse carriage ride to his Davos hotel. He had volunteered to research the spirit of the small alpine town to prepare for the cell's planned protests during the upcoming World Economic Forum summit. He did bring his hiking boots, just in case high-level research would be needed.

Bob checked into his room and started to prepare for the next days' talks with their local allies. They had gained the confidence of a local Internet service provider who had promised to host their net connections during the protests. The cell planned to cover the centre of Davos with a closed Wi-Fi network with repeaters and sufficient redundancy to allow encrypted push-to-talk calling. It was likely that the authorities would monitor short-wave radios and even GSM presence, but with Wi-Fi they had more options to cover themselves.

Now it was time to get to the 'net. That proved to be harder than expected. He had brought his mobile gear, the laptops, the wireless access points, routers, and stuff. However, there was no connectivity in the wall jack. After checking the IP setup, calling reception, calling the hotel manager, summoning their tech maintenance person, and much debugging, he was finally able to ping the cell's home server. However, for whatever reason, the VPN connection never worked. Bob was not enough of a hacker to figure it out.

After a few hours of wasted trying, Bob decided to call it a day. He opened the storm windows, inhaled the distant smell of horse manure, listened to the cowbells, and decided to talk about net connection options with their provider the next day. Over a pot of fondue, he hoped.

The developments outlined above – increased mobility of people, ubiquitous data access, freedom of time and place, fragmentation of activities, multitasking, short attention spans, nomadic use – together define important parameters for the use of mobile devices and, consequently, mobile content. They set requirements that mobile systems must be able to fulfil. For instance, human–machine interfaces designed for mobile use should support *glance-and-click* interaction allowing people to carry out meaningful functions in brief periods of time. Or, since people can create and enjoy content anywhere, anytime, we cannot make too many assumptions on the use situations. Such requirements will be discussed at length in Chapter 6.

2.2   Mobile Device Categories

What makes a device mobile? There are numerous factors that affect mobility, for example, carryability, battery life, and intended tasks.[4] A mobile device should be small and light; it should fit nicely into the user's hand and pocket. It should be easy to carry at all times. Some devices may be worn in arm or wrist bands or carried on neck straps or belt holders.

Mobile devices can be considered as *information appliances* that are computer-enhanced consumer devices dedicated to a restricted cluster of tasks (Bergman 2000). They are notably different from desktop PCs, in that they have limited functionality targeted only to specific tasks, they are not upgradeable, but are perceived to be less expensive and less complicated to use.

We can divide mobile devices into three categories:

1. dedicated media devices;

2. Swiss army knives; and

3. toolbox devices.

We categorize the devices based on the number of functions (or features) and their versatility rather than by shape, form, and wearability or carryability. Among other features, these have a major influence on the user interface design, as described in Chapter 6. Such a categorization also aids in analysing the evolution of mobile devices – when the functionality increases, the devices eventually move from one category to another.

Next, we introduce the characteristics of each category, and provide examples of devices in each of them.

2.2.1 Dedicated Media Devices

As the name of this category implies, a dedicated media device is designed for a single purpose in such a way that its functionality cannot be extended or upgraded further. Such devices are mainly intended for getting (creating) or enjoying *one type of content*. These devices may have additional, secondary features (for example, an MP3 player with a radio receiver, or a digital camera with a display for previewing the images), which support and complement the limited number of primary tasks. A common characteristic for these devices is that the user does not manage content with them.

New functions cannot be added, except by the manufacturer after a new device generation is introduced, so this class cannot be used for

---

[4] For further discussion related to mobility, see http://opengardensblog.futuretext. com/archives/2006/01/

general computation. As a consequence, an external device (often a desktop or laptop computer) is needed for more complex content management tasks, such as updating and archiving content, and composing collections.

An exception in this category is the wide range of devices by Archos[5] that are tailored for storing and enjoying different types of media (audio, images, and video). Similarly, the latest Apple iPod[6] models are multimedia devices handling various types of content. They are versatile devices, yet not upgradeable as far as features are concerned.

Next, we will briefly present some typical devices in this category.

2.2.1.1 Digital Cameras

The transition from film-based photography to digital photography took place practically overnight. It is easy to spot digital cameras by the dozen in festivities, holidays, and even everyday events. Traditional film cameras are quickly becoming curiosities. Today, only die-hard enthusiasts and late adopters stick to film. More and more people now carry digital cameras and shoot anything, anywhere, since individual images cost practically nothing. Others carry cameras only on special occasions, such as weddings. At the time of writing this, digital cameras are becoming standard features in mobile phones, which greatly increase the availability of always-with, always-on imaging. Many cameras and phones also feature video recording.

2.2.1.2 Digital Video Cameras

These devices store the captured video stream either onto a solid state memory card, on an internal hard drive, or directly on a DVD. Most modern digital still cameras double as modest video cameras (and vice versa), reducing the need to carry two separate devices and blurring the line between these two device classes.

2.2.1.3 MP3 Players

Portable music players are quickly becoming commodity items all over the globe. They typically support several compression formats but the iconic abbreviation 'MP3' has been widely adopted to refer to portable digital music files in general. While there are variations on the storage technology (flash cards, hard disks), connections, and battery and UI solutions, the main function in all kinds of mobile players is identical:

---

[5] http://www.archos.com/
[6] http://www.apple.com/ipod/ipod.html

pump music into headsets. More sophisticated players allow managing of playlists (section 2.7), and include functions beyond music use. For instance, Apple's iPods do games, slide shows, contact lists, and even videos. The various iPod accessories have developed into a rich gadget and fashion market of their own. It could be said that the two key functions of an iPod are music enjoyment and personal expression. As a consequence, iPod has gradually transformed into a "Swiss army knife" (see below).

2.2.1.4 DVD Players

DVD players were offered in portable formats first for vehicle use and later for other personal use. The relatively large screen sizes (at least compared to phones and iPods) can offer sufficient viewing quality for nomadic users, but usage while walking is not practical. It seems that dedicated video players are preferable for extended viewing, and smaller devices excel for quick video clips for those gaps in life. The UIs in dedicated players are minimal, which can be sufficient for their intended single purpose.

2.2.1.5 Fitness Monitors

This is yet another class of devices that are becoming multi-purpose gadgets. The original functions, like heartbeat monitoring or step counting, are being supplemented by navigation, compasses, altimeters, and other kinds of sensors. Typical mobile fitness monitors include separate sensing devices, such as chest bands, shoe sensors, and UIs in wrist-worn units. Such devices are best for data gathering and occasional monitoring, but other functions can be cumbersome due to the small size.

Figure 2-4. A mobile GPS navigator with multi-continent maps. Featuring a hard disk, this device still has extra memory space to accommodate MP3 music and digital photos.

2.2.1.6 Navigators

Navigators are becoming common companions for people outside their home bases. The maturity of GPS technology and the availability of accurate maps make current navigators most suitable for use in vehicles and outdoor recreation (section 7.2. provides further discussion on location-based applications and content). High-rise buildings in cities often obscure navigation satellites, so personal navigators may still take a generation or two to fully support pedestrians. Dedicated navigators can offer better usability, but the same functionality is available in a number of phones and PDAs.

2.2.2 Swiss Army Knives

"Swiss army knife" is our nickname for a multi-purpose device that can be used for performing several different tasks *on many content types*. However, extending its functionality to new tasks and content types is still difficult or restricted. For example, many mobile phones provide a limited set of pre-installed applications for voice communication, messaging, imaging, time management, and music. To broaden the application palette, the users may download Java applications to the phone, but they do not integrate fully to the phone's computing platform, resources, and services. Moreover, the device may not have enough available memory to store several applications. The extensibility is there, but it is limited.

Sometimes devices initially designed for a single purpose later include capabilities for other purposes. For example, the mobile phones were at first dedicated tools for voice communication but have later evolved into multifunctional gadgets. Similarly, Apple's iPod was originally designed for listening to music, while current versions may be used for browsing and viewing images and video, as well as managing time and contacts. Interestingly, iPod is *not* designed for obtaining, maintaining, or sharing content, and for that purpose the user must use the iTunes client application on a desktop computer.

Similarly, in handheld gaming consoles, Sony PSP[7] is leading the evolution of this device class towards more feature-loaded devices. When compared to arguably similar Nintendo DS,[8] Sony PSP is closer to our Swiss army knife category, because it is designed also for browsing web, music, video, and images, whereas Nintendo DS is committed to gaming.

---

[7] http://www.yourpsp.com/
[8] http://www.nintendo.com/systemsds

Because Swiss army knives provide a wider range of features than dedicated devices, the UI often contains more physical controls, and possibly larger and better displays. One typical usability limitation is the lack of multitasking, that is, the device can only be used for one function at the time. In contrast, some toolbox devices allow users to switch between several concurrent applications (see below).

From the personal content perspective, the devices typically provide more computing power and storage space, and better wireless and/or cable connectivity. Some devices have embedded cameras, and a possibility for audio recording.

Typical Swiss army knives provide limited possibilities for third-party application development; for instance, the Java platform restricts the application environment to gain enhanced security.

2.2.2.1 E-mail Clients

E-mail clients, such as the Blackberry,[9] are intended primarily for a single task: to handle mobile e-mail. The design is optimized towards this task, which seems to attract large numbers of buyers. The same functions can be carried out with many phones and PDAs, but there is definitely something to be said for single-purpose gadgets.

2.2.2.2 Game Consoles

Handheld game consoles are starting to feature multimedia functions besides games (which are indeed rich multimedia experiences in their own right). The Sony PSP, discussed above, plays movies and music, and can be used to browse the Web upon WLAN coverage. Again, the usability has been optimized towards mobile gaming, but the devices can still be practical for content viewing. Game gadgets have traditionally relied on memory cards as the main means of file transfer, but more devices now include wireless connectivity.

2.2.3 Toolbox Devices

The third category – toolbox – is the next step up the mobile content evolution ladder. Devices in this category are rarely media-only devices, but instead are sometimes referred to as "multimedia computers", with media features as well as many other equally important features (voice communication, messaging, time management). This implies that depending on the designated core tasks and thus the design of the device, its primary function may vary. For example, Nokia ESeries[10]

---

[9] http://www.blackberry.com/
[10] http://www.nokiaforbusiness.com/emea/eseries/

phones are targeted to enterprise use and NSeries[11] phones for multi-media, although both use the same Series 60[12] platform.

The devices, such as PDAs, smart phones, and multimedia computers, are open for adding new functionality, services, and content types. Users may install, for example, new e-mail clients or video players. The devices provide open software platforms for (third-party) application development.

In this class of devices, particular attention must be given to the design of multi-tasking and task switching. For instance, what to do when the user is listening to music and a phone call is coming in? Most likely the music should be paused and the user alerted. If the user accepts the call, how to return to the music once the call ends – should the previous states be automatically restored or not? These and similar issues are discussed in Chapter 6.

For third-party application developers, the toolbox devices are appealing targets, as their platform is open and they have more processing power and memory, better graphics and audio capabilities, embedded cameras, and various connectivity features (such as, GSM, 3G, WLAN, and Bluetooth).

2.2.3.1 Laptop PCs

The most obvious and most versatile mobile computing device, the laptop computer, is a trusty companion for many workers on the move. Even non-corporate people have found that bringing a laptop along gives them the freedom to keep their electronic environment with them. For instance, on a long trip the laptop is needed for uploading images from a camera or for changing the music on an MP3 player. Telltale signs of PC pervasiveness are the hordes of people hunched over their laptops in places with WLAN connectivity, such as Starbuck's.

While a laptop is clearly a viable platform for almost any application, it is notably ill-suited for many tasks that people carry out while on the move. For instance, speaking on the phone, typing a short message, listening to music, or navigating a foreign city are indeed possible on a laptop but are more conveniently done with other devices while on the move.[13]

---

[11] http://www.nokia.com/nseries/index.html

[12] http://www.s60.com/

[13] However, laptops in cars are moderately usable for navigation, entertainment and various mobile office tasks. In general, cars are closer to nomadic use situations than mobile ones, as people may spend extended amounts of time in their vehicles. Still there are frequent interruptions while driving, so mobile situations also apply there.

2.2.3.2 Wearable Computers

Wearables conceal the laptop's shape and size to allow some degree of use while on the move. Wearable computers are always on and always ready, they are totally controlled by the user, and they are considered both by the user and others around to belong to the user's personal space (Mann 1997). There are several possible wearable computer designs, typically comprising head-worn displays, single-hand keyboards, and pointing devices, with battery and computing units mounted in backpacks, jackets, or belts (Figure 2-5). Usually the setup includes wireless links to the Internet, speech input and output, and maybe also some sensors for location, orientation, and physiological parameters.

Despite the capacity and true mobility, wearable computing has so far remained the domain of geeks and specific occupations, such as industrial maintenance workers. In terms of accessible content types and formats, wearables present no major differences to laptops.

2.2.3.3 PDAs

A personal digital assistant is more limited in its interaction capabilities compared to laptops. Interaction typically takes place with a stylus and text is entered with graffiti-like strokes or an on-screen keyboard. Since

Figure 2-5. A wearable computer (Photo: courtesy of Tero Hakala).

PDAs normally do not include keyboards, most users have to stop moving in order to do anything with their devices.

PDAs include most functions found in PCs, but the most popular application appears to relate to PIM: calendars, phonebooks, short notes, voice memos. Many people also do e-mail and Web browsing on PDAs, at least when the connections allow. Mobile music, games, navigation, translation, and e-books are also commonly used.

There are a rich variety of content types that can be viewed and edited on PDAs, and the UI limitations only seem to exclude most graphically or computationally extensive applications. Previously, network connectivity relied on synchronization of PDA data with networked sources while docked at a home base, but nowadays many devices have wireless connectivity features (WLAN, 3G) and the boundary between smart phones and PDAs has started to blur.

2.2.3.4 Smart Phones

Mobile phones were traditionally designed for communication, primarily for talking, but later also for messaging via SMS, instant messages, and e-mail. Modern smart phones include the same PIM functions as found in PDAs, and there are a variety of specialized phone designs for imaging, gaming, fashion, and sports. The most fundamental factor that differentiates phones from common PDAs is their connectedness via cellular data channels. Mobile phones are (potentially) always on, always connected even on the move.

While phones are not yet true Internet citizens, the limitations are more economical than technical. The interaction with mobile phones is not much different to that with PDAs, although phones typically shun pointing devices and favour keyboards that are more usable while walking or driving (not that we advise doing so!). With integrated cameras and music players, modern smart phones now treat images, videos, and music as everyday mobile data types.

2.2.4 Accessories and Other Devices

The three categories described above may have many *accessories*. Here we divide them into the following categories based on their purpose:

- *Improving mobility*. For example, a headset for hands-free voice communication; extending use with high capacity batteries or travel chargers for car lighter or USB port; enhancing wearability or carryability with belt cases, pouches, straps, and armbands;

- *Adding new functionality*: For instance, wayfinding with a Bluetooth GPS receiver;

- *Enhancing interaction*: A portable wireless keyboard for text input; digital pen with a Bluetooth connection for note taking and drawing; acceleration sensors (Samsung S310) for sensing the motion and orientation of the device; launching services and accessing phone functions by touching an RFID tag (Nokia Xpress-on RFID Reader Shell);

- *Enriching media features*: larger memory cards for storage; loud-speakers; printing via Bluetooth; module with Bluetooth connection for showing content stored in the device on TV; adapter for connecting the device to the home stereo;

- *Facilitating stationary use*: For example, car kits for mobile phones enable hands-free use while driving; a car kit for iPod integrates the device to the car audio system; desk stands ease the battery charging and synchronizing; and USB cables can be used for connecting the mobile device to a PC.

As the above listed devices (and next week a plethora of new ones) become more widely adopted, the amount of personal data originating from these devices will explode. Personal content management must be summoned to help. So the glorious march of shiny gadgets is closely followed by less glorious but equally important content management methods – the topic of this book.

2.3   Mobile Compromises

So far we have discussed the trends towards mobility, and the mobile devices we use. Now it is time to turn our focus to the consequences of mobility. What sacrifices, if any, do we need to make while on the move? Is mobile use different from stationary use? How is mobile content different from stationary content?

The shortest answer to these questions is probably compromised quality: less pixels, worse sound, shorter documents, cumbersome interaction. Yet, there are methods that can compensate for loss of quality, and in some cases make mobile content more valued than stationary content.

Some technical and physical realities cannot be dismissed. Mobile devices simply have to be smaller and lighter than their stationary counterparts, otherwise their mobility will suffer.[14] This means that some of their essential functions need to be compromised. Displays are smaller, keypads are tinier, processors are slower, and memories smaller compared to the devices found in living rooms and offices. This does not mean that mobile devices are always doomed to be worse than stationary devices. There will always be tradeoffs of mobile benefits versus mobile limitations, given the intended use.

Power consumption is a key factor for any mobile device. Modern applications call for more processing speed, which comes at the expense of battery life. Faster electronics equals greater power consumption – unless the circuits can also be made significantly smaller in size.

POWER TO THE PEOPLE

Cathy is a self-confessed power junkie. She routinely forgets to bring her mobile phone charger with her on the road. Even a little visit to the City gives her some headache, as she needs to find a place to recharge. She's got a ton of friends in the Greater London area, though, so there's always a place to wind up. Besides, that gives a convenient excuse to meet people.

So there she was again, at Paddington station with only one bar left of the battery. Cathy had talked to Deena in the train very briefly – maybe just an hour – so why was it out of juice already? Steve had explained that with poor reception, like in a train, the battery would die more quickly. Ah, whatever.

Cathy called some friends. Out of luck – Michelle was in Provence, Suzie busy in a session, and Denise could not be reached. What to do now? Switch off the phone? That was such an alien idea. Instead, Cathy decided to bite the bullet, follow Steve's advice, and get an extra battery for her phone.

There was a Phoney store a block away. Cathy walked in, explained her problem, and was out in five minutes with a new battery, complete with new juice. That should make it. Now she would only have to make sure she remembered to bring the second battery with her on the road . . .

The miniaturization of electronics in the past decades has allowed constantly increasing efficiency in terms of computing cycles per watts consumed, and this trend will continue. Improved battery technologies have also allowed higher energy storage densities. However, at the same time, mobile gadgets have got bigger and brighter displays, and mobile applications have become larger in terms of their memory

---

[14] Again, less so in the case of car electronics. Despite being mobile, vehicles can carry devices of substantial weight and size. In this book, the notion of "mobile" focuses on devices that we carry personally.

footprint and computational requirements. Their functions have also become more diverse.

As a result, many mobile devices are heavily compromised by their batteries. Without radical changes in battery technology, such as a breakthrough in fuel cells, this will affect most mobile devices in the years to come.

Consider a mobile smart phone with media capabilities. A typical smart phone can function for a week on a fully charged battery, thanks to various sophisticated energy saving strategies. When the phone is in use for talking, the practical battery life drops to several hours. When the power hungry multimedia functions are engaged, the batteries are drained faster. As a consequence, phone users – not unlike the users of any battery-powered devices – find themselves constantly monitoring the battery levels, planning for charging strategies, optimizing the use of functions in terms of energy consumption, and preparing to recover when the juice runs out. While some of these inconveniences can be helped with better technology, mobile power will always be scarce. Power consciousness is therefore a fundamental defining factor of mobile use.

Connectivity is another factor that heavily affects mobile use. The available connection speeds vary widely. A hotel room may be equipped with a 100 Mbit/s LAN, while low-end cellular data access points give maybe 9600 bit/s – a difference of four orders of magnitude. The users of mobile devices need to be aware of the different connection technologies and their costs and limitations. For instance, nobody should attempt to download a DVD movie over a GPRS connection,

Figure 2-6. A phone research concept by Nokia, enabled by new flexible battery technologies (design: Jari Ijäs, reproduced by permission of © Nokia).

as the eight GB of content could take days to download and cost maybe hundreds of dollars in cell phone fees. Estimating speeds and costs is, of course, hard for non-technical users.

The problem gets worse when some of the data resides in networks and some is available locally in the users' devices. It may not be apparent at all to users if they are accessing data locally or over a network. This should not matter at the end of the day, but once the issues of speed and cost step onto the stage, the users suddenly engage themselves into optimizing data access patterns. Some people go to great lengths with this; others just do not care or simply give up. This is a fundamental problem of mobility: as the technical circumstances vary greatly, it is difficult to hide the resulting technical limitations, as users still need to make choices on the basis of these limitations.

DAVE, THE BIT PINCHER

Dave hated cell phone plans. For one thing, he could not figure out why his data bill tended to skyrocket. He had gone with his grandson Eddie to the local Phoney store and settled for the Mr. Data concept, which supposedly was what he needed to access those betting sites via GPRS, while not spending all his wins on the data costs. Eddie had explained that with this deal he could access a total of 100 megabytes per month, included in the plan. Given his usage, the limit would be more than sufficient. Dave did not have a clear idea of what a GPRS or a megabyte was, but he was happy with the simplicity.

Now, something was broken. The connections seemed much slower than usual and the latest bill was more than £100 higher than expected. Dave had ordered the detailed listing from Phoney and then invited Eddie to come and explain it to him. Eddie had showed up promptly, good boy, with his new girlfriend. She watched the tube and conversed politely with Esther while Eddie played Sherlock on the bill, then the phone, then the bill again, and so on.

After a while it turned out that, as Phoney had sent Dave a new SIM card, the default data connection settings had been reset. Eddie explained that instead of a native data connection he had unknowingly dialled modem calls, which took time and therefore cost money, while being so slow.

Dave felt helpless. How could he have known? Why was new technology so obscure? Give him something that he can hit with a wrench, and he would be happy. If it leaked, Dave could fix it – but this new tech did not just leak, it was totally broken.

Mobility also means that the users' data moves about in various environments. Some of these places can be hostile. For instance, WLAN networks are vulnerable to wireless attacks. Mobile devices with WLAN could be exposed to attacks simply by being within reach of a network with rogue devices. An attacker could lurk in a mall waiting for people with unprotected PDAs, and then try to eavesdrop on the data traffic or implant malicious software into unprotected devices. To avoid such risks, data encryption is a key function in wireless networks including WLAN, GSM, and Bluetooth.

Compromising may take place passively without any malicious attacks. Simply the presence of a mobile device may suffice. For instance, some mobile phones allow access to their sharing media libraries through UPnP (Universal Plug and Play) networking over Bluetooth. Sharing is disabled by default, but could be still left on without the user knowing it. Similarly, an attempt to print an image via Bluetooth in a supermarket leaves a trace of the event in the shopkeeper's devices.

While personal content is not always too risky by its nature, it is nevertheless personal, and therefore merits protection. The Internet is, among other things, an instant massive distribution channel for personal blunders. Consider the story of the "Star Wars Kid", a schoolboy who recorded his vividly enacted light sabre exercises to video that then leaked into the Internet (Star Wars Kid 2003). The video spread instantly over the globe and was seen by millions of people. Had the boy known that his personal moments could spread so widely, he might have not recorded them at all, or kept them carefully to himself.

Sadly, better data protection usually means decreased ease of use. The technologies and use conventions for the security of mobile personal content are just starting to appear, so the compromises between usability and security are still being explored among mobile device designers. Meanwhile, the users are making their own solutions to mobile content security. The consensus could take a long time to develop.

HANDS OFF MY FILES

Angie is quite annoyed. They had come to this party at Joey's, one of Eddie's hacker friends. Besides computers, Joey was into DJing, and had found an interesting way to combine the two. Eddie explained to her how Joey's sets were quite special: he invited people to bring their mobile phones and wireless music players, and then broke into them, digging out the music and playing it to everyone. Eddie seemed to think that was the neatest thing.

Many people seemed to agree. The guys and gals were dancing to the beat with their hands in the air, going wild when a new mix began and someone shouted "That's my song! Howd'ya do that!?" Joey just nodded his head, smiled enigmatically, and kept on clicking his PC while the tune pumped out of the speakers.

> Eddie explained that Joey looked for the open ports in the short-range links that were common in today's devices, and then worked his way into the shared files. Often Joey did not even have to break anywhere, as sharing was on by default in many devices, but sometimes he had to spend over five minutes to uncover the encryption codes.
>
> Angie found it appalling. It was maybe fun to bring your songs into the party, but what about other data? What could this hacker get into? She had a lot of private things on her phone that she wanted to keep to herself. For all she knew, this Joey Hacksta could copy all her data to that DJ laptop.
>
> When Eddie was not looking, Angie switched off her phone. That's a barrier even Joey can't cross, she thought.

2.3.1 Teeny Weeny UIs

Any user of a mobile device is painfully aware of the limitations of the mobile UIs. For instance, most people would probably agree that the displays and keypads in their mobile phones are too small. Yet they are prepared to type and read lengthy text messages with the same awkward UI, as long as the topic of the message is highly valued. Given sufficient interest and patience, we will tolerate almost any limitations in our gadgets. We will compromise quality and convenience for added mobility.

It is difficult to make the keyboards and displays on a phone larger, as that would make the phone larger. On the contrary, customers seem to prefer smaller and thinner phones. The input/output devices of a phone must necessarily reflect compromises between desirability, usability, reliability, and price. The same compromises pester the makers of all mobile devices.

However, there are ways of getting around the size limitations. For instance, while the keypad of the average phone is too small to include the QWERTY layout, a lexicon-based predictive input method (such as T9[15]) can give higher text entry speeds with fewer keys. Speech recognition does away with the keyboard altogether, yet introduces a number of new challenges (Chapter 6). Displays can be shown on virtual head-worn screens (Figure 2-5), or moved to other nearby devices. In short, there are various ways of designing around the limits, but at the same time, new complications are introduced.

2.4   Because it Can!

If there was a single enabler for advances in electronics in the past several decades, it would have to be the continuing trend in miniatur-

---

[15] http://www.t9.com/

ization. The famous "Moore's law" by Gordon E. Moore (1965) estimates that computing capacity (CPU speeds, memory sizes) will double every 1.5 years or so. This astonishing prediction has held true for a long time, largely thanks to the increasing miniaturization of integrated circuits. The diminishing size of transistors allows for larger numbers of elements onto chips and decreased power consumption. As chip makers cram ever more elements on chips, new functions become practical, and old functions become faster – or use less power.

A similar development, thanks to advances in micromechanics and quantum physics, has allowed hard disk manufacturers to build ever denser storage devices. At the turn of the century, hard disk capacity was increasing faster than the Moore's law predicted, due to a leap in magnetic reading head technologies. At the time of writing, the capacity race has slowed down, but mass storage prices keep falling, so the amount of storage per unit of money still follows the Moore's law.

There have been, over time, predictions of Moore's law failing due to various physical barriers. For instance, as microcircuit features sizes decreased below 100 nanometres, the inherent uncertainty of manipulating electrons in near-atomic size circuits was expected to render further improvements in digital electronics impractical. However, advances in solid state physics, circuit design, and IC manufacturing have made the current nanoscale processes an everyday practice.

As another example, the ever increasing number of input/output connections of the chips demand more pins on their packages, which makes device design more difficult. To counter this difficulty and gain yet more speed, major processor makers are increasing the number of CPUs on a single chip to gain performance by multiprocessing, and integrating peripheral devices (such as cache memories) on the same chips.

It appears that the various physical barriers will eventually be circumvented through advances in physics, clever circuit design, and better manufacturing. However, we are not expecting fundamental slowdowns in the advance of electronics in the next decade or two. Technology will march on – *because it can*.

2.5   Convergence

So the trend of better, smaller, cheaper, faster will continue. What will it give us?

DEVICE	FUNCTION	ADD-ON FUNCTIONS
Watch	Keeping time	Alarm, calculator, pulse meter, contact list, compass, altimeter . . .
MP3 player	Listening	Games, calendars, contact lists, photo albums, video . . .
Mobile phone	Calling	Messaging, contact list, web browsing, imaging, gaming, navigation, blogging . . .
Game console	Playing	Messaging, music, video, net browsing, online communities . . .
Calculator	Arithmetic	Programming, graphing, symbolic calculus . . .

Table 2-1. Some mobile devices, their essential functions, and add-on functions.

As we said above, we will get more functions, more speed, more memory size, and more of everything.[16] Mobile devices will continue to add more functions that once belonged to separate gadgets. As evidence, consider Table 2-1. The add-on functions are constantly increasing in number, sometimes surpassing the original intended function, as current smart phones, or multimedia computers, demonstrate.

The term *digital convergence* is often used to signify this trend of including ever more functions in single devices, which leads to a situation where many devices can carry out similar functions.[17] For instance, as memory prices fall, an MP3 player on a chip will become so cheap that it could even become a standard function in any device – including some non-obvious ones, such as light bulbs, wall sockets, batteries, glasses, cigarette lighters, shoes, jackets, or anything. While such combinations may not always make perfect sense, others will. The convergence will go on *because it can*, and new surprising combinations of functionality will keep coming up, with more new forms of digital content (Figure 2-7).

As more devices can carry out an increasing number of functions, and therefore can handle more kinds of content, the typical uses of the devices can become blurred. The devices become *interchangeable*,

---

[16] However, the human physiology is still constant, so we are not getting any faster– better–cheaper in our mental processes. This creates the "human bottleneck" that is a key underlying factor to be considered in the following chapters.

[17] There is a law in computer hacker folklore which states that every piece of software includes more and more functions until its capable of reading e-mail and browsing the Web. We feel there is a clear analogy to digital convergence.

Figure 2-7. Digital convergence – a refrigerator with an integrated panel for viewing recipes, news, weather, even digital photos.

and the content will become more important than the devices used to create and access it. Ultimately, content may move freely between devices, perhaps moving with us as we move about in the world.

2.6   Wireless Revolution

While microelectronics continue to give ever greater computing power to the masses, a more quiet but potentially equally important area of technology has been steadily advancing: communications. The speed of networking has evolved at a somewhat slower pace to microelectronics, but the results of that evolution have affected our lives in perhaps even more fundamental ways.

One below-the-radar change is the inclusion of digital networking capability in many previously unconnected devices – home stereos, cameras, TV recorders, even refrigerators and washing machines. While an Internet-enabled toaster[18] is more of a curiosity than a practical appliance, it still is indicative of the trend towards

---

[18] http://news.bbc.co.uk/1/hi/sci/tech/1264205.stm

the coming convergence. Things will be networked, *because they can*.

As we write this, broadband Internet connections have just been transformed from a novelty into a commodity in typical Western households. Cable and xDSL-based access has just become stable, fast, and cheap enough to entice people to the notion of *always-on* Internet. This is a fundamental change: connecting to the Internet no longer requires waiting for modem dialups and callbacks, so people start to perceive it as an ever-present part of their world that can be summoned at whim. Similar changes in life are happening all the time as communication technology advances.

Arguably, wireless communications has given people more freedom than most other areas of communication technology. The mobile phone has irrevocably transformed our society. The sociologists are still out in the field to study these rapid changes, but even without scholarly proofs we know that our life has changed in a number of ways.

For some of us, life has become a little more relaxed. We no longer need to make elaborate plans to meet people for lunch in a busy city – we just agree to call when we get into town. The Helsinki youth, our mobility benchmark,[19] flock in the city centre in patterns driven by *ad hoc* information they receive from their friends. They do not bother to agree on a place where to meet with their skateboarding buddies, but instead monitor the location of people and get together with the help of text messages and mobile chat.

For others, life has become more hectic. The mobile phone extends the reach of the corporation everywhere. We can be working anywhere, all the time. Some praise this newly found 24/7 freedom; some have become slaves to it. Our society is still adapting to this new ability to be always on.

Mobile phones are currently the most visible form of global wireless connectivity. For personal content, the other important mobile channel must be WLAN. Especially in North America, but also in China and South Korea, WLAN is reaching millions of mobile Internet users in cafés, stores, and on the street. While many of the networks are private and closed, there is an increasing movement towards networks available to general public.

The typical WLAN application is normal Internet use – browsing, e-mail, blogging, chatting – but new uses are emerging. Voice-over-IP clients are now available for laptops, PDAs, and phones. There are

---

[19] Helsinki is one of the first mobile cities where cellular technology truly has affected the behavioural patterns of the youth since the mid-1990s (Kopomaa 2000).

technical areas where cellular has an edge over WLAN (for instance, roaming, charging, and power economy), but for many people WLAN seems to be good enough, at least in urban areas.

In the longer term, as all networks converge towards IP protocols, there will be less differences between the various *access technologies*. WLAN, Wimax, GSM, 3G, and other mobile networks can simply be different ways of connecting to the Internet, the mother of all networks. Well, almost all networks. Except maybe for broadcast TV.

2.6.1   Broadcast Networks

Traditional radio and TV networks were designed for the broadcasting model: one source, millions of viewers. The Internet is inherently a many-to-many network, and not ideally suited for mass distribution of broadcast content. For that, it is more economical to use satellites that beam down DVB (digital video broadcast) content. In existing urban areas it is enticing to use the existing TV cabling and terrestrial transmitters – and even phone lines.

Granted, countless Internet-based casting stations exist, for instance, for the popular Internet radio. However, there the infrastructure is not currently benefiting of the economy of distributing the same content all over the network. Instead, individual users get their own data stream that is not shared with other users.[20] This does not matter with small numbers of users, but becomes an issue when millions subscribe to the same content. However, it is another matter altogether about what will happen to broadcast media in the longer term. It is possible that broadcasting networks will mutate and join the stream of other Internet data.

Mobile TV is now being piloted around the world. Digital TV receivers are being embedded in mobile phones (such as Samsung SGH-910 in Figure 2-8). One of the standards, DVB-H, uses essentially the same MPEG coding technology as satellite TV, but the transmission chain has been modified for mobile terminals. The design allows adaptation into the various quirks of mobile radio links, such as multipath fading and Doppler effects, to name but two. In particular, the content can be economically received with relatively low energy consumption, which, we repeat, is crucial for mobile phones.

As the mobile TV pilots are just starting, mass adoption of this technology can be years away. However, from the consumer's point of

---

[20] Some peer-to-peer designs attempt to change this. It is currently too early to say if those approaches will fly for broadcast TV.

Figure 2-8. A mobile phone with a DVB-H TV receiver (Photo: courtesy of Tero Hakala).

view, the experience is essentially the same as with a home TV. Well, quite a small home TV, and infinitely more portable. However, more functions could be added that benefit the mobility. For instance, context information (see Chapter 4) becomes more relevant to mobile TV. What could be the new show formats that a personal context-aware mobile TV allows? The jury is still out on that question.

2.6.2 Short-Range Wireless

In the short range (within a few metres), the focus of wireless communications has been less in general purpose networking and more in connecting several accessories together. Bluetooth, for instance, was designed to connect a phone with its various accessories (such as wireless hands-free sets), or connecting a wireless mouse or a printer to a PC. Historically, such short-range links have been designed for minor power consumption and thus low data rates. More recent designs, for instance Ultra-Wide Band (UWB) links, promise far increased data rates, while Wibree promises even drastically lower power consumption.

Do the data rates in wireless links really matter? Yes – in a big way. The data throughput capacity of a wireless network (or any data link, in fact) directly dictates how much content it can pump to and from mobile devices. That, in turn, heavily affects the usability of the network. This may not matter for small files such as brief e-mail messages, but becomes crucial with larger files, such as video. People may be prepared to wait for a few seconds to receive an MP3 song, but waiting a day to download a feature movie from YouTube (see section 7.3 on video sharing) is not practical with wireless devices yet.[21]

A typical Bluetooth transceiver found in many mobile phones has a practical data rate in the order of 500 kbit/s. This means that it takes a minute or two to beam an MP3 song from one phone to another. This is a wait many people tolerate. However, copying a DivX movie of 500 MB this way could take hours. For faster data rates, clever college students would probably hook up phones via USB into PCs, or simply swap memory cards.

The channel capacity has, we repeat, a major effect in the usability of media over that channel. Given the various access technologies both in the long and short range, with different capacities and costs, it is not straightforward to decide which channel to use. The notion "any content, anywhere, anytime" is simply not realistic at the moment – and may never be, as long as the communication channels differ in a way that affects the users' convenience or their budgets.

Speed does not come for free, either. It always costs energy to transmit bits, and the faster you send bits, the faster you run out of juice. For larger data transfers, small media devices may need backup power. As a consequence, wireless users need to become somewhat aware of such parameters as transmission times, power consumption, and costs. This is a huge challenge, considering that it is the general public that we are talking about! Moreover, the application designs must take the various channels into account. Even more fundamentally, the variety in access methods has an effect on the design of the content formats and content architectures.

Consider the case of a large, high-quality MP3 collection, which might be, say, 200 gigabytes. It is not practical today to fit that collection into an MP3 player. However, the player could still easily contain the ID3 tags of all those songs and albums. The metadata (see Chapter

---

[21] Will wireless communications become the norm? Is the time at hand when we can break loose of the tangled wires? Sorry, not quite. Wired communications is always faster than wireless by, say, two or three orders of magnitude. When WLAN gets to 500 Mbit/s, the wired Ethernet will already be doing hundreds of gigabits per second. We may be eternally fated to balance between wired speed and wireless freedom.

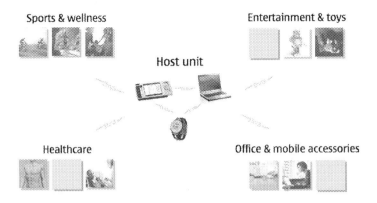

Sports & wellness

Entertainment & toys

Host unit

Healthcare

Office & mobile accessories

Figure 2-9. The Wibree short-range networking technology seeks flexible power consumption / speed tradeoff (reproduced by permission of © Nokia).

4) can be cheaply carried along, even though the actual content sits somewhere at home.

Due to limitations of channel capacities, intermittency, and other factors, a personal content architecture must be designed to live with content that is missing or only partially present. We will discuss these aspects in Chapter 5.

2.7   Case Study: Mobile Music

So far, we have concentrated primarily on mobile devices, and the consequences of us becoming nomadic. Now we will turn our focus closer to content, and look at one particular area of mobile content: music.

In this section, we review the recent history of mobile recorded music and discuss how it has evolved in terms of devices and user control over the content. This area is now undergoing a rapid change from physical to digital artefacts, and therefore serves us well to illustrate the kinds of developments that will follow in many other areas of mobile content. Use patterns related to mobile music are discussed in more detail in section 6.2.

Ever since the birth of gramophones in the 19th century, music listeners have been able to take their favourite music on the road. Portable record players usually offered inferior quality compared to those installed at home, but what mattered to people was the portability. They could enjoy the music at friends', at parties, on the beach, or any other place that had hitherto been the domain of the radio. They were now able to select the music at whim.

Even though singles and EPs had a significant share of the vinyl market, the LP record became the fundamental unit for music distribution. An LP was often designed to be listened to as a whole, ideally from the beginning to the end. Selecting individual songs from an LP required some dexterity. As a result, the LP became a time-continuous collection of strongly related songs, broken in two only by the need to flip the record around.

Tape recorders (also known as magnetophones), popularized in the 1950s, gave the consumers the freedom to choose and order the pieces of music they wanted to listen. People got used to the concept of recording songs from the radio, vinyl, and other tapes. Some old-timers still fondly reminisce on how they used to stand by the tape recorder, fingers ready to hit the record button, waiting for their favourite songs from the radio. However, due to the linear nature of the tape, it was slow to search for a particular recording and difficult to reorganize the contents. The tape and vinyl formats were complementary, both having their uses.

The few commercial portable $\frac{1}{4}$" tape recorders were mostly used by professionals and devoted amateurs. It was not until the later 1960s that portable music recording and playback became practical for the masses, as Philips popularized the C-cassette format. The sound quality was worse due to the smaller tape width ($\frac{1}{8}$") and slower speed, but the decrease in quality was made up for by the small size and practical handling. Various portable C-cassette recorders flooded the market. The essence of the tape technology remained the same, but the size decrease allowed a shift in the use patterns. In turn, this led to greater adoption, larger manufacturing volumes, and decreasing prices. The C-cassette arguably brought personal music management to the hands of masses.

The next fundamental change followed when Sony introduced the Walkman[22] in 1979 (Figure 2-10). Within the next decade, the easily portable cassette player with headphones became the standard armoury for the youth of the "Walkman generation". The basic use of the player remained identical to that of previous tape recorders, but the device was now small enough to be carried literally everywhere. Also, shifting the sound output from speakers to headphones allowed greater individual freedom for listening anywhere, any time. Reaching beyond mere music consumption, the device became a statement of youth values, social escapism, fashion, and a multitude of other factors. The

---

[22] Walkman was unarguably the iPod of the 1980s.

Figure 2-10. Sony Walkman (Photo: courtesy of Tero Hakala).

Walkman thus allowed new use patterns that were to a large degree placed in the social situations of their users.

In the analogue age, the music stored on a device ranged from a few minutes on a gramophone recording to typically 60 minutes on a vinyl recording. A C-cassette contained 60–90 minutes, whereas a $1/4''$ tape could hold several hours of music. With the transition to the digital domain, the amount of carryable music increased several orders of magnitude: MP3 players can easily contain music for several weeks.

The first digital distribution format for consumer markets, the CD (compact disc), was introduced in 1982. Besides offering enhanced sound quality, it improved the random access features of the LP record with digital indexes. People could now easily skip tracks and program playlists with their CD players. However, until recordable CDs became widely available in the late 1990s, people again had to resort to tape to build their own compilations. Mobile CD players were introduced in 1984,[23] targeting the Walkman-style headphone listening. By the turn of the decade, the CD had practically replaced the vinyl disc.[24]

While the CD success story evolved, the Minidisc format was announced by Sony in 1991,[25] for people who either wanted to record on the move or simply wanted smaller media to carry. The portable Minidisc players were physically smaller than CD players and allowed

---

[23] Sony Discman D50, http://www.sony.net/Fun/SH/1-21/h1.html

[24] Yet, vinyl still refuses to die. Some devotees are buying vinyl records, sufficiently to keep the format alive.

[25] http://www.sony.net/Fun/SH/1-21/h4.html

on-the-go editing of the playlists. The technologically inclined consumers were endlessly debating whether the ATRAC compression coding, used to fit an hour of music to a small Minidisc, deteriorated the sound quality too much when compared to a CD. However, in some areas, Japan for instance, Minidiscs were rather popular.

The most popular music compression method so far, the MP3 (more precisely: MPEG-1 layer 3), surfaced from the Fraunhofer laboratories in 1992 and 1993 when the coding algorithm was integrated into MPEG-1, and the MPEG-1 standard was published, respectively. It allowed compressing digital music to 10%–15% of the original data size on a CD, with only minor effects on sound quality. As a result, it became practical to distribute music over the (then) speed-limited Internet, which had meanwhile been spreading like wildfire over the globe. First, technically advanced users would rip songs from CDs for listening on their computers and for sharing with friends on the Internet. Later, mobile players based on digital memory allowed enjoying the MP3 files on the road.

MP3 players allowed users a great degree of control over the song order. The collections of songs had become known as playlists, which are a new form of personal content. The MP3 files not only include music but also metadata relating to it in the form of ID3 tags (section 4.6). The tags typically detail the song name, artist, album name, publication year, and other data. This metadata facilitates digital management of the music files. Today, ID3 tags are the fundamental method for searching and sorting MP3 music.

More fundamentally, MP3 came to represent to the digital generation the same freedom as the C-cassette had represented to the analogue generation. It allowed people to take control of their music (and friends' music too). It brought back the power of home-brew music compilations and playlists that the fixed CD format had suppressed.[26]

By the turn of the century, there had appeared numerous Internet sites for sharing digital music, many of them illegal. For instance, *MP3. com* and *Napster* allowed uploading and downloading of commercial music for free. Peer-to-peer (P2P) file sharing networks (such as *Kazaa*) allowed efficient distribution of music files from the users' computers. For some users, P2P networks became a primary channel for obtaining music. More recently, lawsuits have been posed against music sharing sites, and new legitimate commercial online music sites have appeared.

---

[26] Let us also remember that much of the world is still not thoroughly infested with MP3 players and PCs. The C-cassette is still a very viable format in many parts of the world, for instance in India, Asia, and Africa.

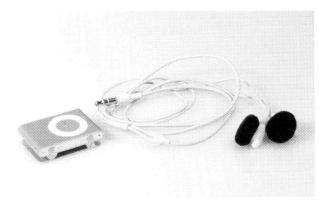

Figure 2-11. The iPod from Apple has become the icon of digital music (Photo: courtesy of Tero Hakala).

Nevertheless, according to one source, downloading in P2P networks seems to increase.[27]

The iPod player series from Apple has today become the iconic representation of mobile music devices (Figure 2-11). While there are a plethora of other mobile players on the market, the iPod defines the essence of mobile music culture in many ways. Apple has coupled the mobile device into their iTunes online music service and the iTunes client application for personal computers. Together, the players and the service allow people to enjoy their music, manage their music collection, search and purchase new music online, and share playlists with other users. With the addition of podcasts and video, the experience is now flowing over to other forms of content.

Since an MP3 song is typically stored in one MP3 file, the single song has become a fundamental unit of digital music, in addition to whole albums. The whole notion of an album is changing, as song collections may now include videos, multimedia art, and remixable versions of the music.[28]

In the case of Apple, music files (and their metadata) are now living in multiple places: iPods, iTunes libraries in PCs, and in the iTunes

---

[27] The number of simultaneously logged P2P network users globally: 2003 – 5.5 million; 2006 – 9.4 million. Source: http://www.p2pnet.net/story/10459

[28] Since digital music (or indeed any content) can be reused easily, remixing has become the norm in some genres, particularly urban music styles. We will talk more about remixing in Chapters 4 and 6.

music store. The store offers background information on the artists and their albums, conveniently linked for random access. It maintains a history of the users' purchases and manages the digital rights related to the downloaded files. In essence, the song that the user enjoys on the road is just an end-product of this sophisticated content management chain. The notion of a song is then, in a way, changing to represent the whole chain, not just the end-product.

As mobile music has transformed via various steps into today's bit collections, users' control over their music has increased with every step.[29] With such freedom, dealing with the music has mutated from physical manipulation into digital information management. Music can now be conveniently handled in the fully digital domain, with the artefact (the song) only residing as a series of bits in a memory device. This has caused some discomfort to people who would prefer to hold music as physical artefacts. Such people often like to burn downloaded music onto CDs (and print the covers), just for being able to touch the artefact and admire cover art without the use of computers.

Other forms of media (video, books, news) are undergoing similar transformations from analogue to digital, creating ever greater needs for content management. People are becoming IT department managers in their digital content warehouses. They are facing the associated digital difficulties – but also the added digital freedom. Our goal is to design content management in such a way that enjoyment is maximized but the need for management is minimized. This book hopes to uncover some ways of achieving just that.

2.8   References

Bergman E. (ed.) (2000) *Information Appliances and Beyond: Interaction design for Consumer Products*. Morgan Kaufmann, San Francisco, CA, USA.

Gleick J. (2000) *FSTR – The acceleration of just about everything*. Knopf Publishing Group.

Kopomaa T. (2000) *The city in your pocket: Birth of the mobile information society*. Gaudeamus, Helsinki, Finland.

Lifton R.J. (1999) *The Protean Self: human resilience in an age of fragmentation*. University of Chicago Press.

---

[29] Consumers' definition of "their content" will differ from that of record companies. As we write this, the notions of fair use, personal ownership, copyrights, and content protection are in a state of turmoil.

Mann S. (1997) An historical account of the "WearComp" and "WearCam" inventions developed for applications in "Personal Imaging". Proceedings of the *First International Symposium on Wearable Computers* (ISWC), 1997, pp. 66–73.

Mitchell W.J. (1999) *E-topia: "Urban life, Jim – but not as we know it."* MIT Press, Cambridge, MA, USA, 184 p.

Moore G.E. (1965) *Cramming more components onto integrated circuits.* Electronics, Volume 38, Number 8, April 19, 1965.

Negroponte N. (1995) *Being Digital.* Knopf, New York, USA, 272 p.

Oulasvirta A., Tamminen S., Roto V., and Kuorelahti J. (2005) Interaction in 4-second bursts: the fragmented nature of attentional resources in mobile HCI. In: *Proceedings of the SIGCHI Conference on Human Factors in Computing Systems* (Portland, Oregon, USA, April 2–7, 2005). CHI '05. ACM Press, New York, NY, USA.

Rosenberg R. (1998) *Computing without Mice and Keyboards: Text and Graphic Input Devices for Mobile Computing.* Doctoral Dissertation, Department of Computer Science, University of London.

Star Wars Kid (2003) *Star Wars Kid Remix Videos Pictures MP3 Video Download Ghyslain Starwars.* Online document available at http://www.jedimaster.net/.

Chapter 3: Mobile Personal Content Uncovered

application **device** digital **file**
information **management**
mobile object photo **user**

So far we have learned that mobility, instantaneous use, and computing regardless of the time and place, are currently prevailing trends in many walks of life. Mobility brings new, spontaneous aspects to our life, changing many patterns of our daily behaviour. Instead of sending a map describing a route to our newly purchased home, we agree with the first-time visitor that they give us a call when exiting the highway, and the instructions are then given in real-time. This decreasing need to pre-plan activities is one of the most characteristic aspects of mobility. It brings along an expectation of *content-on-demand*: getting access to any required information *whenever the information is needed*, instead of only *when it can be obtained*.

Personal Content Experience: Managing Digital Life in the Mobile Age J. Lehikoinen, A. Aaltonen, P. Huuskonen and I. Salminen © 2007 John Wiley & Sons, Ltd

It is one of the primary drivers for the need for mobile content: "I want my content here, now. No need to worry where it has been stored or how I can find it. Just give me what I need, whenever, wherever."

Mobility is not only *accessing* content. An equally important aspect is the *creation* of content; storing important events and experiences by the means of photos, video clips, audio recordings, or textual blog entries. Not to mention the implicit content creation activities, such as creating a heart rate log while jogging, or a dictionary entry for the T9 predictive text-entry method when typing a text message on a smart phone. The mother of all personal devices, the mobile phone that is always on, always with you, is increasingly recording such information. So it is not surprising that digital personal content creation is becoming a common daily activity.

Yet another aspect inherent in mobile personal content is *sharing*. The basic human need for sharing experiences with loved ones (or even with the general public) not only characterizes personal content but also couples it tightly with mobile communications technology.

This chapter discusses mobile personal content, while considering its primary characteristics. Essentially, mobility (the content going with you wherever you go), with its distributed nature opens up new perspectives in understanding the phenomenon of digital personal content. It influences the user experience by providing new opportunities for creating and experiencing the content regardless of the time and place. However, it poses constraints related to small screens, limited input modalities, and so on.

Another point worth considering is that many issues related to mobile personal content are dynamic. The necessary technology standards are still under development, the usage patterns have not settled down yet, and the whole ecosystem is just starting to take shape. Thus, the underlying technologies and behavioural patterns will radically affect the ongoing development in the near future. These aspects are discussed further in Chapters 5 and 6.

There are several ways to approach mobile personal content. A traditional approach is related to a comparison between traditional digital personal content and pointing out similarities and differences. This is a promising approach, since much that is taking place within the domain of mobile personal content has earlier traversed through similar phases within the desktop computing domain. As content that has traditionally been analogue (for instance, photos, and recorded TV shows) is captured and used in digital form, it is also important to compare it with the devices that were used earlier for creating and accessing such content, and devices that will be used when mobility is concerned. Such issues were discussed in Chapter 2.

Yet another aspect is the term "computing" itself. Many digital consumer devices are, technically speaking, computers, but are not considered as such by their users. Instead of mobile computers, they are referred to as their analogue counterparts, such as "music players", "phones", etc. (see section 2.2 on mobile device classification). Therefore, emphasizing computation in mobile devices only blurs the scope.

There are several factors that affect experiencing content (Figure 3-1). As can be seen, content is tightly coupled with many aspects relating to people, technology, and business, and should therefore not be investigated as a separate phenomenon. This figure acts as a rough roadmap to the issues discussed in this book.

In addition to direct actors, there are also a number of indirect actors that affect the content objects, via the direct actors. For instance, prevailing *trends* affect the choices made by users, as well as content creators. Similarly, *society* affects people through *legislation* and providing statutory support for artists. Also *culture history* and *norms* affect content creation and usage. Society also controls businesses. Together, all direct and indirect actors ultimately affect our personal content experience.

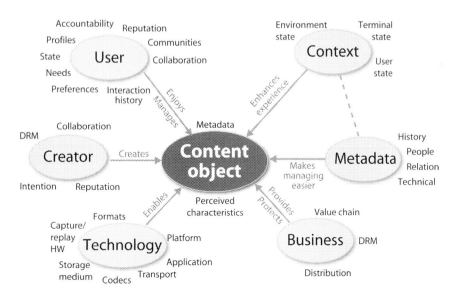

Figure 3-1. Some actors that affect content directly.

3.1   First there were Files

There are probably hundreds of thousands files stored somewhere inside the mass memory of your computing device, be it a desktop computer, a laptop computer, or a mobile phone. Many of these files are required by the operating system or different applications for the device to function properly, and you as a user do not need to be aware of their existence.

Then there are files that are of explicit importance to you. They may be digital photos, e-mail messages, or songs purchased from an online music store. You have probably experienced such files, by listening to a song, or shooting a photo while on holiday. In other words, you already have some kind of relationship with them. And yet, from a computer point of view, a file is just another collection of digital information – bits and bytes tied together, arranged in a specific order, and interpreted in a specific way.

The files are stored in a hierarchical fashion in a tree-like structure by using folders, special containers that may contain either files, or other folders (Figure 3-2). Any file in the hierarchy can then be uniquely addressed by its name and path to the containing folder.

In order to access the files, you have two options: either you use the file management functions offered by the operating system, such as the Windows XP Explorer, or you use a specific application targeted at operating on certain kinds of files only.

The problem in both cases is that regardless of the approach, you, as the user, must have some understanding of the location of the files. You need to know the internal file system organization (the folder hierarchy) together with all the other aspects you should be aware of,

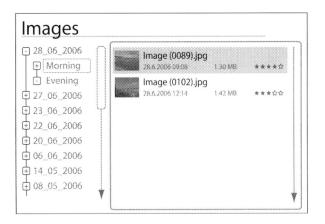

Figure 3-2. A schematic presentation of a file manager.

such as disk drives, networks, etc. Navigating in the folder structure and performing operations on files based on their name and path is referred to as file management.

3.1.1 From File Management to Content Management

Essentially, as far as personal content experience is concerned, *file management* relates to managing files and file structure (that is, creating, copying, moving, and renaming the actual files and directories) by explicit interaction with the file system, regardless of the information that the files contain. *Content management*, on the contrary, includes generic operations – over all content types – that concern the actual contents and interactions with it. Hence, content management includes tasks such as editing, indexing, transcoding, controlling versions, sharing, and searching, regardless of the underlying technology and the way content is stored and accessed (for discussion between file system and content databases, see section 5.2).

As we pointed out in Chapter 1, content management is often considered as a process that information managers and Web content managers need to take care of in large enterprises (Boiko 2005). The key difference is that the content management systems we are addressing in this book are targeted at all of us, including those not interested in technology, and those who hate computing. All of them will eventually have to start using some computing platform for storing and retrieving their personal content. This is the kind of user base we are discussing here.

A step beyond file management, towards managing content instead of files, is providing access to the content based on content type, not by its location, and allowing operations based on the contents of the files themselves (Figure 3.3).

The next step is allowing *virtual containers*. A good example is Picasa 2, an application targeted towards managing photo collections.[1] Picasa 2 allows the user to create photo albums, referred to as labels, which may contain photos from any existing folder. The labels contain links to the actual locations of the photos. Labelling allows the same photo to be included in many collections without having to make explicit copies.[2] Furthermore, it frees the user from some of the burdens related to file management. The application takes care of all extra logic required to implement the virtual containers.

---

[1] http://picasa.google.com/
[2] Links, or shortcuts, are in use in many file systems as well. However, the way they are used and managed is similar to regular files.

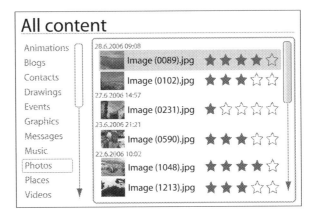

Figure 3-3. A step towards managing content: no folders, but content organized by type and then sorted and grouped by time.

The problem with approaches such as that of Picasa's is that eventually file management will be needed anyway. Since the solutions are application-specific, they are not supported by other applications. Managing the photos within the application is usually easy and self-evident. However, when the photos need to be accessed in some other application, or at the operating system level (such as when making backups), the concepts of labels and folders may become confusing. It may not be explicitly clear where the actual files are, and whether moving a file from one folder to another, for instance, breaks the label structure within the application.

Another implication of the specific nature of these solutions is that it makes switching from one application to another more complicated. It would appear that all carefully crafted application-specific features need to be re-defined from scratch. The consequences of this are yet to be seen, as they will tend to become more severe as time goes by.

EASY COME, EASY GO

A few years ago Cathy had a digital camera. She used it extensively at first, shooting pictures of everything, particularly all the flowers and plants in the neighbourhood. She then uploaded the images to her computer and added keywords to the pictures with the FotoX application her friend recommended. She spent months compiling a database of pictures of different herbs, with unusual labels (i.e., "breaking his nose"). That database would come in handy later when she needed to find a certain plant to describe particular emotions.

Then, as so often happens, the company behind FotoX went out of business. Cathy learned – much to her dismay – that the application ceased to work when the Web-based licensing scheme no longer had any servers to connect to. Her painstakingly collected database was nothing but a pile of incomprehensible bits. The images were there, as files, but the keywords she had entered were apparently not in those files. Even her techie friend Steve could not bring the labels back, despite his array of conversion tools.

Cathy considered re-entering all those keywords again, but never got around to doing so. Instead, she found herself shying away from digital cameras altogether. If digital information was so fragile, it would be better to stick to good old watercolour and gouache . . .

The phenomenon described in the above scenario is familiar to those dealing with preserving digital information (Chen 2003). The fact that technology advances rapidly, resulting in devices, protocols, and file formats that are not backwards compatible, implies that a lot of digital information is lost forever. Content preservation is dealt with by Content Lifecycle Management (CLM). In the past, CLM has been discussed in the context of enterprise knowledge management; see e.g. Päivärinta and Munkvold (2005); however, it is becoming increasingly important in the personal content domain.

With some mobile devices the approach is somewhat different. Most devices, at least those targeted towards other than traditional computing tasks, intentionally hide the file structure from the user in the tasks where it is not explicitly required. As a result, an application such as file manager may never be needed. Typical examples include mobile phones, such as Series 60[3] smart phones. In those cases, the concept of a file is not required in most tasks. The inherent problem is that even though the concept of a file system is not abstracted to the user interface level, the system itself is still based on files. This has many indirect implications for the user, yet they have no means of addressing them. Challenges arise, for instance, when moving files from a PC to the mobile device. Locating the transferred file with the mobile device's user interface may be difficult.

Above, we have briefly discussed a transition from file management towards managing content. What, then, distinguishes mere files from content? Now it is time to dive into some discussion involving definitions and characterisations, so be prepared.

---

[3] http://www.s60.com/

3.1.2 Creation and Usage make Content Personal

According to the trivial definition, content can be characterized as: *Content is everything that is inside a container.*

This is straightforward because if the mass memory is the container, then data stored inside it is content. In this sense, content is a broader term, which also includes data – data being an instance of content. However, in the context of this book, we take a different view to content.

In multimedia and computing jargon, we tend to associate content with information closer to the human senses. For example, photos taken by you are considered content, whereas initialization files stored in binary format usually are not. As a consequence, we will start by characterizing content as:

Content is data that is targeted at human access.[4]

Thus, the key question in characterizing content is, "Who is primarily using the data?" If the data is mainly used by the computer for its internal operation, it is not content. If it is primarily used by a person, it is content. This is the predominant factor that distinguishes a file from a piece of content.

From this we learn one significant fact: whether a data object is content or not depends on the person dealing with the content. What is content to an administrator is mere data for an office worker. Therefore, the definition is ultimately always subjective.

We are still not happy with this definition. First, even though firewall logs are data to most of us, they are targeted at human access. This is because they are in a textual form, and their only purpose is to allow the administrator to check what is going on when, for instance, the network once again slows down dramatically. One additional aspect needs to be included in our definition:

Content must be somehow meaningful to the person accessing it.

Second, not all content is stored as single data objects. Several objects together can form a piece of content (for example, a playlist), or an object may be composed of numerous single objects (for example, a multimedia presentation with video, audio, and images). So, the next iteration of our definition of content is:

---

[4] Our definition is in line with Boiko's (2005, p. 4) definition, which states that "[...] content is information that retains its human meaning and context." Furthermore, the Merriam–Webster online dictionary defines content as "the principal substance (as written matter, illustrations, or music) offered by a world wide web site" (http://www.m-w.com/)

Content is data that is targeted at human access, including individual data objects and combinations and collections thereof. It is meaningful to the person dealing with it.

It is obvious that as long as the definition is subjective, the term cannot be defined unambiguously. However, for the purposes of this book, the present definition suffices.[5]

What immediately follows from the above is the term *content object*, which refers to a single piece of content, such as a song, or a combination, such as a photo slideshow with audio commentary.

We still need to define what makes content personal. Put another way: "Do you have personal content stored in your computer? What makes your content personal?" What is personal to you may not be personal to your best friend. Once again, some characterization is in order.

Your relationship to the content does seem important. Perhaps we could define personal content as:

Personal content is content that the user has a relationship with.

The question now is: what is a relationship? We expect that some interaction has always taken place with content before it becomes personal, be it shooting a photo or rating a song. In other words, personal content has been experienced, and the term "relationship" reflects the interactions and experience associated with a piece of personal content. One could conclude that content becomes personal through creation or use.

An obvious flaw is that the above definition does not consider whether you are actually controlling the content or not, which would also make your friend's MP3 collection your personal content, should you happen to like it. Thus the definition needs to be amended, for example:

Personal content is content that the user has a relationship with, and is in possession of or has control over.

For instance, consider your favourite TV show. You watch the show, and at the same time you capture it with your personal video recorder (PVR). Once the show is over, you own a persistent copy of it. Obviously you are able to control the copy stored in the hard drive of your PVR. But what about the video stream at the time of broadcasting? According to our definition, it is not personal since you are not able to control it.

We still need one more addition. You may have a relationship with the document that your employer expects you to finish today. What we think is an essential part of personal content is the voluntary nature of the relationship. To complete our characterization, we state that:

---

[5] See also (Boiko 2005, p. 5) for more discussion on the definition of the term.

Personal content is content that the user has a voluntary relationship with, and is in possession of or has control over.[6]

Now, with this characterization, we can start discussing personal content in detail.

3.2   Categorization

In order to understand the nature of personal content, one needs to understand more of its inherent features. What is it like? How does it behave? What are the mechanisms used to manipulate it? How do personal content items relate to each other? How do they relate to external entities?

A common way of describing a phenomenon is to first categorize it. Categorization is a strong tool for understanding and managing any complex phenomenon.

An obvious, intuitive categorization criterion is *by content modality*. Basically all current human-accessible digital content types can be classified into three main categories: the content is primarily visual, auditory, or tactile. In the future, we may also have olfactory and gustatory content.

However, the above classification is superfluous, as it does not clarify the characteristics of personal content. For instance, how should one classify saved games? Saved games provide parameters for recreating a particular context of a virtual world, complete with the environment, characters, and objects (section 7.4). It is meant entirely for human access, but not to any particular senses. Mostly, it results in a visual presentation of the world; additionally, it may also contain auditory elements. Above all, though, it contains information that enables us to experience the virtual world with our senses and imagination.

Another way to classify personal content types is *by their internal representation*, that is, the way the bits are to be interpreted:

* text: SMS, e-mail, e-books, chat session logs;
* images: photos, drawings, scanned paintings;
* audio: music, ring tones, earcons;
* video: movies, clips.

With this kind of classification, however, we will quickly run out of higher-level terms; for any content type arranged differently from the

---

[6] And then again, when your partner asks you to take yet another photo of yet another fabulous sunset, the voluntary nature of the content experience is to be doubted.

basic types (like saved games), we would need to include an additional category, such as *other*, which would then include all types that do not fit within the basic types:

- other: saved games, skins, bookmarks, presence information, game characters, etc.

This is content that is not in textual (since it is not in the text category) but in binary format. The content may still consist of alphanumeric characters only (such as bookmarks); however, they are not meant to be read directly. The way the bits are to be interpreted is not obvious and cannot be determined by the category. Obviously, an alternative is to add a new category for each new content type. This would not be categorization, however, but a plain list of all potential content types.

The categorization is technical by nature, as bits and internal representations are involved. A more human approach would be to look at the list above and consider the function of the content: is it meant to be read, viewed, or listened to? This might allow us to include new categories: is it meant to be played (game-related personal content would be included here); is it meant to convey your personal tastes (skins, presence information)? This might be a fruitful approach, so let us try to re-formulate the list:

- for reading: SMS, e-mail, e-book, chat session log;
- for viewing: photos, scanned paintings, videos, game movies;
- for listening: music, ring tones, earcons;
- for playing: saved games, game characters, high scores;
- for expressing personality: skins, profiles, presence information;
- for analyzing: personal health information;
- for navigation: bookmarks, maps, landmarks.

Categorizing content *by its function* takes us further, since the importance or meaningfulness of content without function is to be doubted. However, there are still borderline cases, such as with bookmarks used for navigating. However, they also provide access to content with any function. What is important here is that bookmarks are your personal pointers to (in many cases) non-personal content.

It is important to note that the categories described above form only a small subset of all available content functions. Obviously, any action that can be performed with a computing device can potentially contain respective personal content types.

Even with this classification, we should be aware that the categories are not unambiguous. Any piece of content may belong in several classes, such as an e-mail may be in the reading class at the same time as in a communication class. Nevertheless, there is a category for each piece of content.

In addition to classifying content by internal representation or function, there are several additional approaches. We will discuss some alternatives briefly. Many of them are better suited for categorization needs in special circumstances, not for general classification.

Personal content types can be classified by their *maturity*. Is the content type declining (such as voice mail); is it mainstream (such as short messages are at the time of writing); is it emerging (such as multimedia messaging); or is it merely a weak signal of a potential new content type in the future (such as context data is at the time of writing)? This kind of classification may be useful in some cases, for example, when developing new personal content services that need to scale up for future needs.

The content can be classified *by its creation*. Is it created by the user; has it been created together with others; or has it been created totally by others? A need for this kind of classification may emerge, for instance, when dealing with copyright issues. Related to creation is also classification by automation – has the content object been created explicitly by the user, such as a saved game, or implicitly, such as a heart rate log? The problem often related to implicitly created content is its second-class citizenship since no explicit effort has been put into the content creation as such. However, losing a log of all sporting activities for the last three years, for instance, could be considered catastrophic.

At least some personal content can be defined *by its expected lifespan*. Is it targeted at one-time enjoyment, such as a rapidly spreading funny e-mail message on a Friday afternoon, or is it a piece of memory that should live forever? This aspect needs to be considered, for example, when designing management systems. Some content needs to be accessible instantly, whereas sometimes it is enough to know that a certain piece of content exists and it can be retrieved when needed. In a similar manner, the need for backups and archiving can be considered.

The list of content classification schemes given above is certainly not exhaustive. On the contrary, new schemes can and should be generated when needed.

3.3   Characteristics of Personal Content

What does personal content look like? This section aims at presenting some of its typical characteristics. The purpose is not only to provide an understanding of personal content features that affect us all, but also to bring up some fundamental topics that need to be addressed when personal content management systems are being developed.

Perhaps the most important feature of personal content is its incredible growth rate. The amount of digital photos, for example, is growing at an enormous rate: according to one study, it is estimated that from 2004 to 2009, the number of digital photos taken increases by 24% each year (IDC 2006) (Figure 3-4). And this estimate includes only digital photos, not the multitude of other content types.

There are two obvious reasons for the explosive growth:

1. New content is continuously created.

2. Existing analogue content is converted into digital form.

A typical example of the first category is taking a photo with a digital camera. Good examples in the second category are scanning old printed photos or digitizing vinyl LPs.

Two more reasons, not as obvious, can be identified:

3. New content types are born.

4. Existing digital content is transformed into another format.

An example of the third category is a playlist (a file that contains links to the songs that together form the playlist), which was not a widely used content type a decade ago. Taking a screenshot of a computer

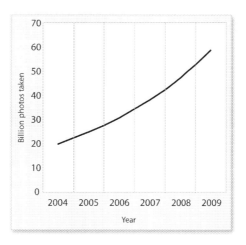

Figure 3-4. Estimated digital photo growth rate from 2004 to 2009 (IDC 2006). The values on the y-axis are only indicative, since the estimated number of photos taken in 2004 varies wildly between 20 and 235 billion, depending on the source.

game, or capturing still images off a video recording, are typical examples of the fourth category.

There is yet another, perhaps even more influential, although less explicit, aspect:

5. More powerful methods of sharing have emerged.

Sharing means sending, forwarding, or publishing on a blog, or making it available on a peer-to-peer network. Essentially it can also be understood as copying. Nevertheless, sharing, together with its strong associations to social behaviour, is one of the key activities in researching personal content, and also typical of many personal content types.

In addition to the features described above, there are three obvious fundamental enablers that have made the content explosion possible in the first place. These include, in no specific order:

- the continuous increase in mass memory capacities;

- the rapid commoditization of consumer broadband network access; and

- active development of more efficient content encoding and decoding algorithms (codecs).

What, then, distinguishes mobile personal content from personal content? In section 2.3, we claimed that a key difference is in content quality, which is often related to the technical limitations inherent in mobile use. What other differences there are? In this section, we will consider the characteristics of personal content in general, and mobile personal content is then discussed in section 3.4.

3.3.1 Content Explosion

The above discussion about content explosion leads us to the first characteristics typical of personal content – its growth rate. It will heavily influence both system design of personal content management system and experience; for instance, Apple's iPhoto 6 is prepared to manage 250 000 photos.[7] To put this in perspective, the mean number of captured digital images per month in the US was 75 in year 2005 (IDC 2005). With such a rate, it would take nearly 300 years for iPhoto's limits to be met . . .

---

[7] http://www.apple.com/ilife/iphoto/

As discussed in section 3.1, in traditional file management the user is expected to have some understanding of the file locations in the folder structure. Indeed, with desktop computers a common practice is to carefully design a folder structure, and then store the files in the folders where they best fit. With the explosive personal content growth rate we are currently facing, this will soon become impossible; it simply requires too much effort to classify and properly distribute thousands of photos over folders, in their different versions and editions, not to speak of all other forms of personal content. Sooner or later, the users will need to choose a preferable way of organizing and managing personal content, that is, transfer from file management to content management.

Related to content explosion is the origin of personal information. Above, sharing was identified an important factor contributing to the growth rate. In addition, another aspect can now be considered. As discussed in section 3.2, personal content may be created either explicitly by the user (such as shooting a photo or capturing a video clip), or implicitly by the application, as a side effect of other user actions (for instance, an automatically saved game, or a suggested route in wayfinding). Often, implicitly created personal content is equally important, even though it may never be directly accessed.

3.3.2 Personal Content is Invaluable

Important and valuable personal belongings have always been kept safe by the means of safes, safe deposits at banks, etc. Digital content is, or should be, no exception. When discussing the features of individual content objects, probably one of the most important of them is, not surprisingly, personality.

In many cases, content has not become personal for free: in order to create or personalize it, you may have had to put in some effort in doing so. If the content is for some reason lost or damaged, the changes made by you are also lost; there is no authority that can restore it. As a consequence, personal content is in many cases priceless: the photo you took at the top of Mt. Fuji may be irrecoverable. The solution is storing the content, more or less frequently, to an alternative storage medium. Backups are discussed in Chapter 5.

3.3.3 Personal Content is Familiar . . . or Not

Yet another essential point is that beyond mere management, users should also be able to *enjoy* their personal content. A proper management system that requires as little user intervention as possible is a necessity but not a sufficient prerequisite for providing positive experiences with personal content. An important aspect is providing *fresh views* to personal content.

One could safely assume that the users know their personal content, at least some of it, at least to some extent. Consider photos, for instance. Knowing the content does not imply that we are able to list by heart all photos taken last year, but rather that it is easy to recognize certain features in the photos (obviously, the same holds for other content types as well): when you see a small thumbnail image of a photo, you may have a fairly good understanding of the event or person the photo presents, even though you cannot make out the details in the thumbnail. In other words, in many cases personal content is familiar and will become more familiar by use.

However, what users seldom know is the way in which the different content objects relate to each other, as discussed in Section 3.6.6. Revealing unexpected, fresh views to content, which were once familiar, can increase the perceived satisfaction.

RELATING TO MADEIRA

Despite her bad experiences with the persistence of metadata, Cathy still kept all her old images of plants, people, and places, such as the trip to Madeira. She did not bring her camera, but luckily her friend Deena did. It was such an amazing island! All those flowers, fern, moss, shrubs. She could have spent years just taking in the flora. Deena, a nature buff herself, dutifully documented all the subjects of Cathy's attention and made photo disks of the images for both of them.

Cathy would occasionally browse these images on her computer in search for inspiration, looking for plants that would convey particular moods, and then use those pictures as inspiration for her paintings. She found that instead of keywords, it was usually ok to just browse. Sometimes she asked the computer to show "related" pictures as a slide show. That was fun! The resulting unusual combinations were sometimes quite enlightening. Or, sometimes, just plain weird. What can you expect from machines, after all?

These so-called *wow* effects can, at best, be the primary driver for rapid and thorough user adaptation to personal content management systems. Discovering associations between pieces of content and presenting these associations to users is a good start towards providing an enjoyable personal content experience. Auto-summarizing content is another potential technique that could be applied. The intimacy associated with the system is increased when it appears to know the user's past, sometimes even better than the user remembers it themselves.

3.3.4 Favourites

Not all personal content is equally valued. Some songs are listened to more than others, and some photos are viewed more often. There are favourites among the piles of personal content.

An important tool related to favourites is rating. Not only should one be able to specifically mark the pieces of content that are special, but means for rating should also be provided. Ratings can be manual or automatic. Automatic rating can be based on several technologies, depending on the content type, the context of use, etc.

In rating, two aspects can be identified. One may provide personal ratings or the ratings may be public, for instances, provided by a community. Sharing ratings adds to the social aspects of personal content.

3.3.5 Sharing and Communities

Even though the term "personal" implies an intimate relationship between a content object and the user, it is a common practice to share content regardless of its type. There are several levels of sharing, ranging from sending a photo to a close friend to publishing a moblog on the Web. As the mobile phones slowly turn into mobile multimedia computers, introducing increasingly advanced content creation capabilities while retaining the "always-connected" nature, the role of sharing is boosted even further.

There are several motives for sharing. In many cases, it is just sharing a personal experience with loved ones, while sometimes it is a question of seeking for fame (sharing a saved game, for instance). Sometimes sharing may be a matter of competition, such as when sharing a heart rate log in a Web community for others to evaluate. Nevertheless, many personal content objects continue their lives beyond their creator's hard drive.

Sharing is tightly related to the social nature of personal content. As witnessed in section 7.3, many online communities are built around sharing content in different forms. Such communities routinely allow uploading one's own content and linking it with other people and content objects, as well as downloading, commenting, and rating other users' content.

3.3.6 Relations and Associations

It is uncommon for a piece of personal content to appear in a void, without relating to any other piece of personal content. Many photos are related to other photos, taken on the same occasion; most songs belong in a collection (be it a CD or a playlist); messages are necessarily related to at least contact information, and so forth. By automatically discovering relations between content objects, the user is not only able to see the larger picture they belong to, but also gain some understanding of their meaning.

Finding and maintaining relations can be useful in several ways. Relations are a strong tool to facilitate searching. Should the system have some understanding of the semantic relationships between the content items, searches by example, for instance, may be greatly enhanced. Second, analyzing relations and presenting relations to the user unexpectedly may provide pleasant surprises and strengthen the wow effect: "How did my device know THAT?" (section 3.3.3). Relations can be considered the glue that bonds the pieces of personal content together and provides meaning to them.

The true power of relations, however, is not evident until they cross the content type boundaries and tie together all content objects that have a common character. Then we can start enjoying and re-living experiences, instead of just single content objects.

What are particularly important are relations between content objects and people. Since it is personal content we are dealing with, the ability to link our digital memories with our social network plays a central role. Relations between content objects and people further contribute to the community aspect of personal content.

3.3.7 Privacy and Security Requirements

Usually there are different privacy requirements for personal content. As discussed above, some pieces of content you may wish to keep for your eyes only, whereas some you may wish to share with your family or friends. In this case, the access is still restricted, as many different sub-levels or groups may need to be defined. Further on, some pieces of content may be public, so that anyone can access them. Another aspect to consider is that privacy definitions given for a certain piece of content may change over the course of time.

Personal content still concerns us all. As more of our personal stuff – pieces of our life – become digital, more and more computer-illiterate users begin to use personal content management systems. It is of utmost importance to being able to secure the personal content, without any domain-specific skills, so that the general public do not have to worry about whether their contents are currently being stolen or abused, or that people who are unaware of such threats, can remain doing so without catastrophic results.

Metadata (data about data) and its inclusion to shared content is another potential threat to privacy. To avoid distributing unwanted metadata, the privacy levels need to affect metadata as well. Metadata is discussed in more detail in Chapter 4.

In the context of mobile personal content, the importance of privacy and security increases further. The devices, and therefore the content stored in and accessible with them, makes them vulnerable to threats such as theft, overseeing, and overhearing.

For additional discussion related to personal content and privacy, refer to Karat et al. (2006).

3.4   Mobile Personal Content

So far we have discussed personal content in general, mobile or otherwise. Now we focus our attention on *mobile* personal content. The first and most important question is: What is the difference between personal content and mobile personal content? As discussed in section 2.3, a typical difference is quality; due to limitations in device capabilities and wireless network throughput, mobile content objects are often of lesser quality. However, the basic features of personal content are not different in a mobile context. The content can be classified as in section 3.2, and the basic characteristics are similar. After all, it is just jumbles of bits and bytes that we are talking about. However, there are also differences, some of which are more indirect in nature.

3.4.1 Mobile Personal Content is Distributed

Personal content, especially that which is mobile, resides in many different places. It is created in your phone, downloaded to your home PC, or recorded with your PVR. Some of your personal content may be at your workplace, as many of us have pictures of spouses and children in the office. One of the most characterizing features of digital content is how easily it is copied and shared across devices and networks. In technical terms, we are talking about *distributed* content. In computing terms, if something is distributed, it means there is no one location where you can find it. You cannot point to one device and say, "See, son, that's were my content is."

Distribution causes many problems for managing your own content: where to find it, how to make sure all important content is safe, how to manage copies, and other issues. Mobility brings the problems and benefits of distributed content management to the surface. With mobile devices, the content is not only created in more places, but also accessed from more places.

For distributed content there are some unique problems that need to be solved. First you have to *find* where the content is stored or where to store your own. Next you may need to *transfer* the content between devices, perhaps also *transcode* or *adapt* it in order for it to fit some more limited devices. Third, there arises the problem of *interoperable* devices – how different devices and applications can together provide a valuable service to the consumer. Also its not only devices that must work with each other, but different applications in those devices must share a common language and common presentation of the data in order to successfully exchange information. All these aspects are discussed in Chapter 5.

3.4.2 Mobile Content is Tied to Creation and Usage Context

Content created while on the move is inherently related to the context, or situation, of its creation (see section 4.7 for more discussion on context). When such a content object is later enjoyed, the experience is enhanced with some of the captured context. Context also ties the content object to a larger picture. Similarly, while enjoying the content, the context at that time is used to fine-tune the experience. The captured context can also be used further to allow a rich set of related automatic grouping schemes based on location.

3.4.3 The Same Content Types, New Usage Patterns

Currently, there are two mainstream mobile content types over the others: music and photos. There are a large number of other mobile content types, but the ubiquity of these two makes them good candidates for generalization when content types common to both stationary and mobile use are being considered.

What is typical of both music and photos is that the actual file types and formats are essentially similar regardless of the mobility aspect – the same JPEG files and MP3 songs are accessed in desktop and mobile devices alike. However, the way they are used, and the operations performed on them, varies significantly between stationary and mobile use scenarios. For instance, the importance of a song sitting in a desktop computer's mass memory may be determined by the play count, whereas in a dedicated MP3 player with a limited amount of memory, the time the song has remained in the device without being replaced with another song may also indicate the song's importance.

Likewise, contact information is content that is available to most e-mail clients. The way contact information is treated with mobile devices, especially mobile phones, is much more fundamental though: one could argue that in many cases, the user's whole social life is represented by the phonebook in their mobile phone. Furthermore, the contacts are generally considered the most valuable information in mobile phones. This suggests that the importance of a traditional content type, contact, has significantly increased when it is used daily and relied heavily upon in most daily mediated communications.

3.4.4 Totally New Content Types, or Extended Use of Existing Content Types

Mobility and the related use patterns brings along new kinds of information, which results in two issues dealing with content types: either new use patterns require totally new file formats, or existing file formats are modified towards better filling the needs implied by mobility. Nevertheless, the trend contributes to the ever-increasing amount of content and variety of content types.

The most characterizing aspect of mobility is, by definition, dynamic location. Mobile use is not tied to a certain location. The device can be, at least in theory, accessed anytime, anywhere. It is always with the user. These characteristics not only change our usage patterns (discussed in section 6.2 in more detail) but also reveal new forms of content altogether.

Since the location of the user and their personal devices is not fixed, information related to current context, such as location, is an obvious example of the content that has not existed on a stationary PC, where the location remains more or less static. This is with the exception of laptops and other portable computers, as these devices are not very useable while on the move. See section 2.1 for further discussion on static, nomadic, and mobile use situations.

Personalized location-dependent information is typical of mobile use only. Such content includes your own location, the location of your friends, location-dependent personal reminders, and location-based services, such as advertisements. Applications that deal with location-based personal content are discussed in section 7.2.

Another example is presence (or availability) information, which is traditionally used in instant messaging applications (International Society for Presence Research 2006). In traditional PC environments, a presence indicator and its associated text merely conveys whether you are available at your PC, busy working, or away from your PC.

Mobile presence[8] not only conveys your availability, but also your location. This information was not needed when working on a stationary PC.

Ring tones are a good example of new uses for existing content types. Music files have obviously existed prior to the mobile age, yet downloading simple melodies or snippets of songs for use as an incoming call indicator has revealed a novel content type.[9]

A mobile device is usually very personal and close to our body. Information that many mobile devices may potentially collect is data related to our health and physical condition, such as heart rate. It does not usually make much sense to continuously monitor the heart rate while sitting on a desktop computer, whereas when mobile, it is often desired.

An interesting yet not obvious newly-derived content type is a personal dictionary. Everyone knows the pain of writing any text with the mobile phone's keypad. However, T9 text entry system has brought some help by applying language-dependent word prediction.[10] However, the dictionary that is used to predict the word is not perfect: different users have different preferences, many domain-specific words are missing, not to speak of all the cool phrases used by the youth and different subcultures. Therefore, the ability to augment the dictionary with user's own words is an essential feature. And the words added in the dictionary are extremely personal in content, even though usually accessible only indirectly while typing a message.

3.4.5 New Behavioural Patterns

Besides new uses related to content types, new ways to deal with content have and will emerge. A typical behavioural factor is related to content creation. Being able to create content while on the move not only contributes to the enormous content growth rate, but also makes users create content on a more relaxed and spontaneous basis. This, in turn, results in storing information that might have otherwise never been captured and retained. One could claim that mobile creation and access is to personal content what writing once was for

---

[8] An example of mobile presence attributes is provided by Wireless Village initiative that is part of Open Mobile Alliance (OMA). See: http://www.openmobilealliance. org/tech/affiliates/wv/wv_pa_v1.0.pdf

[9] One could argue that in the PC world, song snippets have been assigned to events in a similar fashion for more than a decade. Compared to the whole business models built around selling ring tones, however, it is a minor issue.

[10] http://www.t9.com/

passing information between people and generations. Instead of having to pass information orally, it could be written down and then passed on passively and asynchronously. Likewise, capturing everyday events and experiences contributes to storing and possibly passing on the micro-history of the day.

Interaction with personal content is instantaneous by its nature. For instance, upon capturing a video clip of an unexpected event, the ability to instantaneously share the occasion, instead of having to wait until a PC with a broadband fixed line connection is available, greatly influences the desire to do so. Indeed, continuous connectivity has a remarkable effect on how content is treated, and when. Connectivity has boosted social phenomena such as citizen journalism (Scott 2005), which makes individual passive witnesses of an event also its active reporters.

3.4.6 New Challenges

The way content is used in the mobile domain requires addressing several technical, societal, and cultural challenges that are not otherwise present. The fact that the content not only sits firmly inside the user's home or in a tangible artefact, such as a CD, and that the user expects the ability to access and enjoy the content regardless of the time and place (distribution), causes a myriad of new technical challenges that need to be addressed. These included synchronization, content adaptation, and privacy protection (these topics are discussed in Chapter 5). In a similar fashion, challenges in developing user interfaces for mobile personal content, while taking into account the limited input and output capabilities and in many cases non-optimal use situation, need to be addressed (Chapter 6). Furthermore, new societal norms need to be adopted. A typical example is denying the use of digital cameras embedded in mobile phones in places such as dressing rooms, schools, or restaurants.

3.5   Content Wants to be Free?

The Net interprets censorship as damage, and routes around it.
John Gilmore, Electronic Frontier Foundation

The often cited quote above could be easily extended to any system allowing content creation and sharing, by replacing the word "censorship" with "copyright protection". Not surprisingly, then, Digital Rights Management (DRM), is an increasingly hot issue that concerns personal content directly. It is a matter of determining who can enjoy and distribute content in digital form. It is technology applied by commercial content providers wishing to ensure that the use of copyrighted content can be controlled, and it mainly concerns music, DVD movies, and games.

The increased use of distributed computer networks, known as peer-to-peer networks, has boosted the dissemination of copyrighted material without permission. Rights holders have struck back, and the result is a number of trials against network users. An inherent dilemma is that what is good for the rights owner is in many cases not good for the consumer.

There are several sad stories related to DRM, such as Sony BMG's rootkit case in 2005–2006 (Roush 2006). This is how the story goes.

In order to prevent the listeners from making illegal copies of the music they had purchased, Sony BMG included DRM software in the music CDs. Upon inserting the disk in the CD drive of a PC, this software was installed on the user's hard drive. The software acted as a rootkit, a tool that prevents the operating system from seeing certain kinds of software. Then, without the user knowing, several pieces of additional software were installed in the computer. They prevented the user from making more than three copies of a CD. Furthermore, every time the disk was inserted, the software contacted Sony (without informing the user), reporting that a CD was to be played. Even worse, the installation of a rootkit unintentionally allowed other malicious software to sneak into the user's computer without the antivirus scanner's knowledge. Once the DRM system was discovered, a huge Internet campaign was raised against Sony BMG and its spying software.

Indeed, it is in the interest of media houses to secure their investments by applying copyright protection schemes. However, sometimes they may overreact.

Electronic Frontier Foundation(EFF),[11] founded in 1990, is a body that fights for consumer rights in the digital world. EFF defends free speech, fair use, and privacy. They have successfully fought for the consumer and technology developer rights in courts, for instance, in peer-to-peer network trials. The foundation is targeting especially major entertainment companies, mainly in North America, and seeks, among other tasks, to find alternative ways to making P2P pay artists.

We do not take a stand here on how DRM should be taken care of, nor do we advocate any particular approach over the others. What is imperative here is to ensure sufficient protection for content without sacrificing the experience. For discussion related to copyright and censorship in game-related content, refer to section 7.4.

---

[11] http://www.eff.org/

CENTRALLY CONTROLLED DRM, OR: WHY DIVX FAILED

In 1998, a new scheme for viewing movies was presented in the US, named Digital Video Express (DivX).[12] The popular DivX video codec[13] was ironically named after this, containing initially a smiley postfix: DivX;). The idea behind DivX was simple: the user purchased the movie on a disk as usual, at a significantly lower fee compared to the same movie on DVD. The disk was to be viewed on a dedicated player that was equipped with a modem and hooked up to a telephone line. The first time the user inserted the disk, the player downloaded codes from a remote server that unlocked the protection mechanisms present on the disk. Then, the user was able to play the disk for as many times as they wanted within the next 48 hours or, by paying extra, for a longer period of time, or by paying even more, without any restrictions; then the disk was effectively equal to a DVD. A disk was unplayable unless the centrally controlled rights management system had approved the viewing. And the approval was required each time the disk was to be played. Statistics on the user's viewing habits were also collected.

There were many problems associated with DivX, of which one of the most important was the dependency to the DivX company itself. Should the company cease to exist (as the case actually turned out to be), all disks purchased by all customers were useless. Furthermore, as a consumer, you were not able to take the disk to your friend's home for a nice get-together evening, unless you had purchased permissions to view the disk also on your friend's player. Furthermore, even the price for a disk was lower than that of a DVD, it was still higher than the DVD rental price.

DivX is a good example of centrally controlled DRM, where the users do not know what they are paying for, since the rules may change at any time. It is also an interesting example of personally owned content, for which you may not own control rights. No wonder it did not get acceptance within the target consumer base.

There are alternatives to heavy DRM and spying on consumers. One approach is to innovate alternative reward models, such as Creative Common's "Some rights reserved" approach.[14] There are many licensing schemes available, though all of them require attribution: "You let others copy, distribute, display, and perform your copyrighted work – and derivative works based upon it – but only if they give credit the way you request."

Means such as Creative Commons may be especially suitable for amateurs, while viral marketing may be the way to go for commercial markets (Helm 2000). Viral marketing describes any strategy that

---

[12] http://www.dvdjournal.com/extra/divx.html
[13] http://www.divx.com/
[14] http://www.creativecommons.org/

encourages individuals to pass on a marketing message to others, creating the potential for exponential growth in the message's exposure and influence. Like viruses, such strategies take advantage of rapid multiplication to explode the message to potentially millions.

Considering digital content, *superdistribution* is a potential viral marketing tool (Mori 1990). Essentially, superdistribution is a concept where content may be distributed freely, but it is protected from further modifications and modes of usage not authorized by its vendor unless such rights are purchased. Originally superdistribution was envisioned for software deployment, but the method can be easily extended to any digital content. Combined with a peer-to-peer network and mobile micropayments, concepts such as superdistribution may heavily change the way we trade content.

At the end of the day, one realizes that there is no such thing as free content. In the end, someone has to pay for it. The payment may come in the form of money, or the need to put up with advertising, or having to shift through loads of useless content to find some gold nuggets. Ultimately, someone has to pay for the transmission and storage, even if the content itself is free. Every time bits are being stored or transmitted, energy is being consumed, and that is never free.[15]

We are in the middle of a situation where traditional content industries (music, TV, and film, etc.) are struggling as digital distribution is changing their business. They have evolved their content distribution chains for decades on the basis of physical distribution of physical copies of their precious material. Suddenly, any kid on the 'net can get it for free; maybe illegally, or semi-legally, but for all practical purposes for nothing or almost nothing. There are always associated costs that can be less than obvious. As much as the authors of this book would like, as consumers, get tons of free high-quality content, they are not great believers in the free-for-all model.

The advocates of free content often take the view that personal content can be as good as commercially produced content. That can certainly be true in some cases. There are millions of musicians, for instance, who have never received wide attention, while it can be argued that some the music we hear on the Top 20 of commercial channels is low quality. Or is it? Why is it that somehow the artists made it to the top with poor music? Is it their smashing looks, or their rock star lifestyles, or those fabulous cars and jetliners? Or is it just marketing and productization? Mass iconization?

There are countless factors that contribute to stardom. There are many strategies in getting to the top. Pouring in tons of money to

---

[15] You cannot fool the laws of thermodynamics!

advance a career of a no-name artist is one way. Picking up an unrecognized artist with unique talent is another. For every Top 20 artist there are hundreds of Top 200 artists, and millions of Top 2 000 000 artists, and so on. At the end of the tail, there are millions of people creating good stuff, but it is only enjoyed by maybe a handful of other people, or even the "audience of one".

The same effect manifests itself in the rosters of megabuck content houses. For every megastar they have to support dozens of less famous artists, and nurture hundreds of promising talents, and keep track of thousands of wannabees. Being able to select the widely appealing content and invest into making it a success is their central competence, and a reason for their existence. Even if the content creation channels change, in the end it remains costly to maintain the machinery needed to make megastars. For this reason, we consider it likely that the content industry will in the future still be dominated by a handful of major labels. And, by the way, they will still get their money from us consumers. Because they have the machinery to craft widely appealing content, they will receive our money.

Thus, the emphasis of non-mainstream content creation and distribution is known as the *Long Tail* phenomenon (Anderson 2004). And there is indeed life at the end of the tail. It is often pointed out that the Internet is a great equalizer for those lesser-known content creators. We are seeing many artists and bands who have recently gained fame from humble beginnings, first contributing free music to the Web, and later selling their content commercially.

It seems evident that there is space in all parts of the long tail: in the high end dominated by megastars and blockbuster movies; in the middle populated by locally famous artists and not-so-successful independent movies; and in the tail, zillions of bedroom musicians and amateur home video makers. There is room to succeed in all these places. However, the measures for success, the mechanisms for monetizing, distribution, and everything may be radically different in the different parts of the long tail.

The notion of Web 2.0 (O'Reilly 2005) has been much discussed in the past few years. This loosely defined term is often used to refer to applications that are based on the Internet, use Internet technologies as the main application and UI platform, enable the creation of social networks, and gain significant power out of them. A significant aspect of so-called Web 2.0 applications is that they target not only mainstream content, such as Hollywood movies or Madonna songs, but also cater for the lesser known talent that reaches far smaller audiences – that is, the long tail. Clearly this is near to our topic, personal content.

Usually content that is created for personal purposes is definitely not targeted for large audiences. Home movies of kids' birthdays are created for personal use, maybe shown to friends and sent to Grandma. Garage bands, on the other hand, may only be famous for a handful of people but can suddenly gain significant audiences through the Internet.

Sometimes the content that we have created for personal purposes may still be interesting to larger audiences, either directly or indirectly. This may happen if the content itself is directly interesting to many people, such as a clever remix of a pop video. Indirect interest may come about in cases where the content is captured for personal use but also benefits others through the network effect. For instance, when the first user feeds a brand new CD into her ripping software, she has to fill the song names. The rest of the users then receive the names via a Web service, such as freeDB.[16] The first user provides additional information that directly benefits others. She also benefits indirectly in the form of song names that other people have entered for other CDs.

The Web 2.0 movement is creating various kinds for applications for the management of personally contributed data. We believe many of these applications will be ultimately beneficial for personal content management in the sense we have discussed in this book; however, in general, the application wave has just started to move in that direction.

To conclude, personal content should ideally be kept clear from any rights management and payment issues, unless the users so desire. Or, if such mechanisms are used, they should be made easy enough to be handled transparently for most tasks. The Creative Commons licensing scheme, as described above, comes closest to this ideal of the technologies we have seen so far. It allows expression of rights from the creators' point of view with simple terms, and once a rights scheme has been chosen, it may be automatically used in subsequent similar works.

3.6   GEMS, a Tool for Modelling Personal Content Experience

Throughout this chapter, we have learned that personal content is a phenomenon with implications to many areas of technology. What we

---

[16] http://www.freedb.org/

still need to consider is the human perspective: how do users interact with personal content, and what are the actions performed on it? In this section, we take a look at understanding personal content from this important point of view.

In section 3.2, we discussed different ways of categorizing personal content. A potential categorization model was based on function, that is, how the content was to be experienced. A straightforward way of analyzing the potential interactions with personal content, is by listing these functions:

- reading

- viewing

- listening to

- analyzing

- expressing personality, etc.

The functions above are certainly interactions performed on personal content. However, rather than showing interactions that can be performed on it, they are functions that are associated with a specific content type. How about interactions generic across all these categories?

Our aim is not to create a list of 500-plus verbs containing all potential actions that can be performed on personal content, but rather to provide a tool for understanding the human aspects of experiencing it. A tool that could be used, for instance, as a broad framework for analyzing and assessing devices, systems, applications, and user interfaces dealing with personal content, and providing some insight into categorizing them.

An inherent characteristic of personal content experience is the need to share (section 3.3). Not only is sharing a means to experience personal content with others, but an efficient distribution channel as well. Therefore, sharing can be considered an important action performed on personal content.

Before any piece of content can be shared, it must have been retrieved from somewhere. Thinking about *inputs* and *outputs* logically leads us to a lifecycle-like model, where content comes from a source, and at least some of it leaves in one way or another. Perhaps we could consider these two aspects in more detail:

1. Incoming content: downloading, receiving, capturing, trading, etc.;

2. Outgoing content: selling, sending, pushing, publishing, etc.

Now we get content from somewhere by some means, and later may share some of it by some other means. Then what do we do with the content between these two phases? Now it might be time to re-consider the list resulted in categorization by function:

3. Enjoying content: viewing, listening to, reading, personalizing.

Now we have a three-phase model, where content comes in, is enjoyed, and some of it may be later shared.

Are these all potential high-level actions that we can perform on personal content? For the purposes of maintenance tasks, we may need to add yet another category to the model:

4. Maintaining content: organizing, synchronizing, archiving.

These four phases – Get, Enjoy, Maintain, Share – finally lead us to the GEMS model (Figure 3-5), first introduced by Salminen et al. (2005). The verbs in each category (G, E, M, S) provide some examples of the actions that belong to the category. As an example, editing is considered enjoyment, while trimming (such as cutting unwanted parts out of a video clip) is closer to maintenance in nature. Nevertheless, as often is with issues dealing with personal content, these are ulti-mately subjective.

GEMS describes the lifecycle of personal content usage. First the user obtains the content from somewhere (receives, creates, rips, steals . . .), then they enjoy the content while at the same time main-taining it. Finally, some pieces of content will be shared by sending,

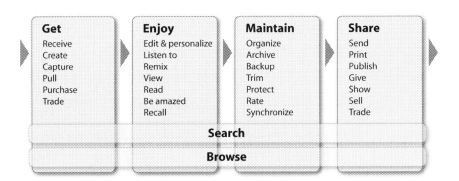

Figure 3-5. GEMS personal content lifecycle model.

publishing, selling, and so forth. Then the shared content object will continue its life in some other user's GEMS cycle. Naturally, not all pieces of content go through all the steps. Likewise, a piece of content may jump from any category to any other, such as after maintaining it can be enjoyed.

Not surprisingly, our model of personal content lifecycle shares many common aspects with the digital information lifecycle model presented by Gilliland-Swetland (2006). In this model, the phases include Creation and Multiversioning, Organization, Searching and Retrieval, Using, and Preservation and Disposition. What is the largest difference is our consumer- and experience-oriented point of view and the fact that the GEMS phases are independent of each other, except for the obvious fact that the "G" phase must precede any other. Thus, some devices may function primarily in the G section (digital still cameras), while others are targeted at "E" (MP3 players). For instance, many mobile devices have evolved over time from mere "E" devices first to cover "S" actions by the means of connectivity, and are now becoming "GES" devices.

Speaking of evolution, in section 2.2 we introduced a three-phase classification for mobile devices: dedicated devices, Swiss-army-knife-devices, and toolbox devices. To some extent, these categories map with the GEMS phases, especially relating to the device evolution from one category to another, where an evolutionary step often introduces new GEMS actions or even a new phase (such as adding connectivity to a device enables "S" actions).

A model as generic as GEMS can never be complete or unambiguous. However, the main purpose of GEMS is to provide a filter through which any system dealing with mobile personal content can be viewed and further analyzed on a high abstraction level. GEMS can be understood as an analysis tool.

Since GEMS is a high-level framework, it can be applied in many ways, specializing when necessary. The simplest cases include using it as a plain checklist: have all the important aspects of available user activities been taken into consideration? More advanced uses include considering an application concept, and mapping its planned functions and services on a GEMS map. This kind of analysis may help in classifying applications and services for finding optimal target user groups. We discuss this approach in Chapter 7, when application designs related to personal content are reviewed and analyzed using GEMS. Chapter 6 presents a more detailed discussion on different GEMS phases related to user interfaces.

One could argue that there is yet another essential action related to personal content that should be regarded an action category all of its own. This action is *searching*. However, searching can also be considered as an action that concerns all GEMS phases. Then, it could be included in the diagram as a horizontal action that spans across all other phases. In a similar fashion, browsing is an action that is needed in all phases (Figure 3-5).

What is missing in GEMS is communication, that is, communication between people. In many cases, communication is a special case of the combination of "G" and "S" phases, such as speaking and listening, but due to its specific nature it could be considered an additional dimension in GEMS. In effect, communication may span across all GEMS phases.

What is interesting to note is that not only content, but also its associated metadata, travels through the same GEMS phases along with its host content object. Therefore, GEMS can also be applied to metadata in understanding the activities taking place when it is used.[17] Metadata and its role in personal content management will be discussed next.

3.7   References

Anderson C. (2004) The Long Tail. *Wired,* Issue 12.10, October 2004. Available online at http://www.wired.com/wired/archive/12.10/tail.html

Boiko, B. (2005) *Content Management Bible* (2nd edn). Wiley Publishing, Inc. Indianapolis, IN, USA

Chen, Su-Shing (2003) Preserving Digital Records and the Life Cycle of Information, *Advances in Computers, Vol. 57, Information Repositories.* Marvin Zelkowitz (ed.), pp. 70–109. Academic Press, MD, USA

Gilliland-Swetland A.J. (2006) Setting the Stage. In: *Introduction to Metadata: Pathways to digital information.* Online edition, v2.1 Available at http://www.getty.edu/research/conducting_research/standards/intrometadata/setting.html

Helm S. (2000) Viral Marketing – Establishing Customer Relationships by "Word-of-mouse." *Electronic Markets*, Volume 10, Number 3 / July 1, 2000, pp. 158–161. Routledge, part of the Taylor & Francis Group

IDC (2005) While Number of Digital Images Captured Increases Rapidly, Consumers' Printing Behavior Changes, IDC Reveals. Press release available at http://www.idc.com/getdoc.jsp?containerId=prUS00220705

IDC (2006) IDC Predicts Increased Number of Digital Pictures Snapped by Affluent PC Users Will Boost Global Photo Printing. Press release available at http://www.idc.com/getdoc.jsp?containerId=prUS20035406

---

[17] A lifecycle model for multimedia metadata has been also proposed by Kosch *et al.* (2005), yet not from the user interaction point of view.

International Society for Presence Research (2006) Online document, available at http://www.temple.edu/ispr

Karat C-M., Brodie C., and Karat J. (2006) Usable Privacy and Security for Personal Information Management. *ACM Communications*, January 2006, Volume 49, Number 1, pp. 56–57. ACM Press, New York, NY, USA.

Kosch H. et al. (2005) The life cycle of multimedia metadata. *IEEE Multimedia*, Volume.12, Number 1, pp. 8 86, January–March 2005.

Mori, R. (1990) Superdistribution: The Concept and the Architecture. The Transactions of the IEICE, Volume. 73, Number 7 (July l990), Special Issue on Cryptography and Information Security.

O'Reilly T. (2005) *What Is Web 2.0: Design Patterns and Business Models for the Next Generation of Software*. Online document, available at http://www.oreilly.com/pub/a/oreilly/tim/news/2005/09/30/what-is-web-20.html

Päivärinta, T. and Munkvold, B.E. (2005) Enterprise Content Management: An Integrated Perspective on Information Management. *Proceedings of the 38th Annual Hawaii International Conference on System Sciences* (HICSS '05), pp. 96–96

Roush W. (2006). Inside the Spyware Scandal. *MIT Technology Review*, May/June 2006, pp. 49–57.

Salminen I., Lehikoinen J., and Huuskonen P. (2005) Developing an Extensible Mobile Metadata Ontology. *Proceedings of the Ninth IASTED International Conference on Software Engineering and Applications*, pp. 266–272. ACTA Press.

Scott B. (2005) A Contemporary History of Digital Journalism. *Television & New Media*, Volume 6, Number. 1, pp. 89–126, SAGE Publications

Chapter 4: Metadata Magic

application		context	data	file	format
image	**information**		management	media	
metadata	object		people		
	standard			user	

In fact, the role of metadata will become so vital as our personal digital repositories grow that metadata's value greatly will surpass that of the content.

John R. Smith (Tešić 2005)

Apparently, several technologies need to be in place before a content management system can function efficiently. The user needs to receive the content, then store it somewhere, and be able to access and enjoy it. All the GEMS model actions, discussed in the preceding chapter, need different technologies, all equally important.

However, we need to raise one technology enabler above the others: *metadata* – data that talks about other data. Metadata is imperative in personal content management. It makes the whole content experience lifecycle manageable, helping to deal with content in

Personal Content Experience: Managing Digital Life in the Mobile Age   J. Lehikoinen, A. Aaltonen, P. Huuskonen and I. Salminen © 2007 John Wiley & Sons, Ltd

fundamentally better ways.[18] For a brief introduction to metadata, see Baca (2000).

This chapter introduces metadata through several concrete examples, and discusses its relation to mobility and an inherently related topic, context-awareness. We especially emphasize the consumers, as opposed to the professional content producers often assumed in metadata-related texts. We present the benefits metadata brings to the users, and analyse potential threats of its excessive usage. We also briefly investigate existing metadata solutions, and discuss some important technologies that are needed to support efficient metadata processing. Finally, we discuss interoperability and related efforts in this area, by considering both devices and applications.

4.1 Metadata for Consumers: A Brief Introduction

The modern world is said to be in the middle of an information explosion, the data available on the Internet increasing exponentially. However, this not new, as we have been there before – as early as the 15th century.

When the technology for printing text *en masse* became widely available, new ways were needed to manage the explosive growth in the available information. In the days of Gutenberg, the information was stored in a physical form. Today the content is digital, but many of the ways for organizing text-oriented material remain the same.

The *library* institution was born out of the need to manage large physical collections of information. To effectively manage a large collection of books, libraries have developed an access system for storing the books in a way that enables people to browse them (Figure 4-1). In some cases, browsing book spines is fine, but if you wish to find a certain book quickly you need a better technique, for instance, some kind of classification.

In a typical library, you will see bookshelves laid out according to some organizational principle. The shelves will show markings that indicate the category of the books contained, and also you will notice that the books in the categories are organized based on a certain sort key. The classification mechanism is used to index books in order to

---

[18] Others also consider metadata important. Try typing "metadata is * important" (including quotes) in google and witness the wide variety of interest areas that benefit from metadata.

Figure 4-1. SYSTEM: library – USER INTERFACE: bookshelf – CATEGORIZATION: subject matter – SORT KEY: author name.

aid in organizing and facilitate their management. Moreover, the classification gives us some hints to the actual contents of the book.

When talking about digital content, the data that is used for describing the content of the object is often called *metadata*. Briefly stated, metadata is data about data and it can be considered as a surrogate or representative of the actual content, thus providing a content summary. Metadata is often understood as text describing the content, keywords known as attributes, and their values. For example, for an MP3 file the metadata attribute "Duration" and its value "00:04:07" tells us that the playing length of the song is a little over four minutes (Figure 4-2).

As the definition implies, the division between data and metadata is not unambiguous. What is data from a certain point of view is metadata from another. For example, a Web service that lists forthcoming TV programmes treats the starting and ending times of the individual programmes as metadata. However, when that schedule is loaded to a PVR for automated recording, the same piece of information becomes input data to the recorder. Similarly, the "Call log" application in the mobile phone stores the communication history (dialed, received and missed calls, and the associated timestamps). The "Phonebook" application may then use this information as a descriptor for a contact. Therefore, the communication information is data for the "Call log", and metadata for the "Phonebook".

Figure 4-2. The metadata attributes of an MP3 file.

4.1.1 Metadata Semantics

For us, the meaning (semantics) of metadata is often evident. The tag "*Quantum physics*" in the classification of a book tells us that the volume is not going to be easy bedtime reading. Computers, however, lack the cultural background and common sense that humans have, and thus need explicit symbolic definition of the meaning of metadata. Much of the work in metadata revolves around this topic: trying to encode its meaning for computers in simple but powerful ways.

The meaning of metadata can be more complex than it would first appear. It is evident that the creator of a painting is its painter, as usually that form of art involves a single creator. It is less evident for other works where multiple people are involved in the creative process. For instance, who is the creator of a blockbuster film – the producer, the director, or perhaps the main actor? The issue gets more complicated when the object that the metadata describes is a part of another object: a song that belongs to the soundtrack of a movie, or the content object is a photo of a famous painting. The importance of semantics will be discussed further in section 4.4.

A fundamental mechanism, critical for navigation and use of content, is a *catalogue*. Metadata is used as the common vocabulary to frame information that we seek to use and these metadata elements are stored and served through a user-accessible catalogue.

Related to personal content, *folksonomies* are of specific interest (Mathes 2004). A folksonomy consists of collaboratively generated, open-ended labels that categorize content, such as Web pages, online photos, and Web links. A folksonomy is a taxonomy where the authors of the labelling system are also the main users, and sometimes creators, of the content. The labels are referred to as *tags* and the labelling process is *tagging*.

In general, metadata describes the document's content, context, or structure (Kennedy and Schauder 1998). Although metadata often comes in a textual or numerical format, it should not be restricted to these, since some objects can be described far more efficiently with other content formats. For example, a thumbnail image embedded into a JPEG image file is clearly metadata since it provides a summary of the actual content and acts as a surrogate when, for instance, browsing several photos.

4.1.2   Metadata – For Managing or Enjoying?

Traditionally, metadata has been the business of content management professionals, such as researchers in information retrieval (IR), librarians, and news archivists, among others. From this point of view, there are several ways to categorize metadata, such as that of Gilliland-Swetland (2000):

- *Administrative* metadata that is used for managing information resources, such as copyright;

- *Descriptive* metadata that describes or identifies information resources, such as vocabularies and annotations;

- *Preservation* metadata that enables information preservation management, such as physical condition of resources;

- *Technical* metadata that illustrates how a system functions or metadata behaves, such as formats and compression information;

- *Use* metadata that describes level and type of information use, such as user tracking.

The above list is typical of the professional content management viewpoint: analytic, technical, neutral, functional. However, iPod-toting youngsters are not interested in the metadata in their players, as long as it does the magic and helps them to find the "good stuff". They might choose altogether different categories for metadata, such as the following for describing music:

- *Genre* metadata that describes the place of a song in the ultra-fragmented field of contemporary music, helping style-conscious listeners to maintain continuity;

- *Energy* metadata that describes the power of the song to make the listeners move;

- *Appraisal* metadata that gives the music figures of merit, for instance by their positions in download charts;

- *Hipness* metadata that gives measures of the song's current popularity and desirability among the consumers; this might differ from the appraisal;

- *Social* metadata that links listeners to other people enjoying similar kinds of music.

If we had to choose between these stereotyped "analytic" and "consumer" viewpoints, we would prefer the consumer view. We would like to achieve systems that allow *content enjoyment*, as opposed to content management. Of course, the analytic and technical views to metadata are needed in the machine room, but the users might prefer to see just the more abstracted and domain-oriented metadata.[19]

Usually people are not well versed in information management, and are unable or unwilling to invest into the mental effort of classifying content into meaningful categories. We merely want either to receive our metadata *pret-a-porter* or have ways for our computers (and other devices) to fill in the metadata for us.

Next we will discuss the two essential challenges related to all metadata management:

1. *metadata creation* (who creates it, how, and when?); and

2. *metadata maintenance* (checking its validity, updating metadata values when necessary, analyzing, and further processing metadata).

4.2   Metadata Creation

Where does metadata come from? Somewhere, someone must have entered it. In the case of personal content, typically the owners of the content fill in their metadata. In the case of commercial content,

---

[19] Of course, there are us tinkerers who enjoy taking things apart to see how they work. There will be people who spend hours tweaking metadata parameters to fine-tune some aspect of their content systems.

metadata is usually defined in content production houses. In addition, some metadata can be filled in automatically without human intervention.

The effort that is invested into entering metadata for a piece content is often inversely proportional to its *uniqueness*. We use uniqueness as one measure of the return-of-investment analysis for metadata.

The uniqueness of an object's metadata is related to whether the object is self-created or available to a wide audience. Commercial content usually comes in abundant copies. There are numerous reasons for this, such as the nature of our economical system (mass production lowers prices and boosts sales), and digital content is fast and easy to duplicate and distribute. To date, commercial interest has always implied large-scale duplication and distribution.

On the other hand, photos are usually unique. Even if two photographers are taking photos at a birthday party, it is highly unlikely that they shoot identical photos. As a consequence, the images feature differences in orientation, lighting, people, objects – even if the photographers take the images exactly at the same time and same place. So even when the physical circumstances match exactly, the contents cannot be the same due to variations in camera electronics and optics, which will render the resulting binary information different from each other. This makes every image unique. Of course, the personal values attached to an image further distinguish it from other similar images.

In essence, each self-created content object exists in small numbers, is valuable only to a few people, and thus there are often only a handful of people who can really say something meaningful about such content. On the other hand, the world is full of duplicates of commercial content objects, which implies that there is a larger pool of people that can describe the content.[20] Individual people can then benefit from others' metadata. The other users may gather bits and pieces of metadata from here and there as somebody else has already entered it. In addition, everybody may contribute more metadata (section 4.1.1). Naturally, this kind of sharing of distributed metadata entry will suffer not only unintentional metadata disintegration, but also mischief by people that abuse or try otherwise to spread false data intentionally (we will return to this phenomenon in section 4.8).

---

[20] However, this necessarily introduces noise in the metadata, as different people may enter similar metadata in different ways. Witness artist and song names in P2P file sharing networks: there must be 50 ways to spell "Paul Simon".

BOB & THE GANG GO TO CHAMONIX

Bob's activist cell is home to many kinds of people. Love for nature somehow seems to be a common trait in these people – most everybody goes backpacking, farms sheep, plants forest, that kind of stuff. So it's only, erm, natural, that during the years they have gone out for all those nature outings, when not busy protesting against megacorporations.

The Chamonix trip last September was a new experience. Not only was it Bob's first hike in Haute Savoie, but this time they had decided to record the trip using all possible electronic means. So they got into it in a big way. Everyone brought digital cameras, some had laptops with them (hardly essential mountain gear, Bob still muses), Joe was recording the spacious alpine ambience into his Minidisc, and DV video was all over the place. They kept an e-mail diary with text and multi-media messages from their phones. A travel blog of sorts, once everything was brought together and organized per time and location.

Bob borrowed a GPS navigator from Ritchie under the condition that he would geotag all his mountain images. Ritch would have come himself, as he's been hiking the Alps dozens of times, but his newborn daughter now gave him deeper experiences during those sleepless nights. The cell guys all promised to capture everything for Ritchie's later enjoyment – well, save that smell of cow dung. Ritch felt that it was smelly enough in their house already what with all that diaper stuff.

A large number of personal content objects are created or otherwise acquired ("G" actions in GEMS) with hand-held, mobile devices. The reason why manual metadata entry is considered hard is partially related to the poor text input capabilities of mobile devices as well as other interaction limitations (Chapter 6). The entry problem worsens when the user is moving (which is, by definition, the norm with mobile devices). The amount of content objects that should be annotated matters too: you may annotate one object carefully, but not necessarily one hundred objects.

Manual metadata is potentially extremely valuable, as the users typically enter things that are important to them, as well as adding semantics to the content. For example, instead of describing that a photo contains two people, the user would probably enter their names. This makes management of data more interesting, as the objects have more meaning. In a way, this signifies a change from machine-readable data to human-readable information.

Even though metadata is generally considered beneficial, it is not to be expected that users will input large amounts of it. It is practical to assume that for most of the consumers *manually entered metadata* equals *non-existing metadata*, even if some users are more active than others. This is because when the amount of content increases, automation of the entry becomes crucial. As a general rule, from the user's perspective, metadata creation should be as automatic as possible.

Automatic metadata is added when the object is created and could illustrate, for example, the technical attributes of the system, its state, as well as technical attributes of the object. Figure 4-3 shows an example of some of the EXIF metadata attributes and values of an image in JPEG format.

File: E:\PICS\IMG_8699.JPG, Make: Canon,   Model: Canon EOS 300D DIGITAL, Orientation: Top left, XResolution: 180, YResolution: 180, ResolutionUnit: Inch, DateTime: 2006:11:05 18:45:52, YCbCrPositioning: Centered, ExifOffset: 196, ExposureTime: 2 seconds, FNumber: 5.60, ISOSpeedRatings: 400, ExifVersion: 0221, DateTimeOriginal: 2006:11:05 18:45:52, DateTimeDigitized: 2006:11:05 18:45:52, ComponentsConfiguration: YCbCr, CompressedBitsPerPixel: 3 (bits/pixel), ShutterSpeedValue: 2 seconds, ApertureValue: F 5.60, ExposureBiasValue: 0.00, MaxApertureValue: F 3.50, MeteringMode: Multi-segment, Flash: Not fired, compulsory flash mode, FocalLength: 22 mm, UserComment: , FlashPixVersion: 0100, ColorSpace: sRGB, ExifImageWidth: 3072, ExifImageHeight: 2048, InteroperabilityOffset: 2366, FocalPlaneXResolution: 3443.95, FocalPlaneYResolution: 3442.02, FocalPlaneResolutionUnit: Inch, SensingMethod: One-chip color area sensor, FileSource: DSC: Digital still camera, CustomRendered: Normal process, ExposureMode: Auto, WhiteBalance: Auto, SceneCaptureType: Landscape, Maker Note (Vendor):, Macro mode: Off, Self timer: Off, Quality: Fine, Flash mode: Auto + red-eye reduction, Sequence mode: Single or Timer, Focus mode: One-Shot, Image size: Large, Easy shooting mode: Landscape, Digital zoom: None, Contrast: High, +1, Saturation: High, +1, Sharpness: High, +1, ISO Value: 400, Metering mode: Evaluative, Focus type: Auto, AF point selected:, Exposure mode: Easy shooting, Focal length: 18: 55 mm (1 mm), Flash activity: Not fired, Flash details:, Focus mode 2: 65535, White Balance: Auto, Sequence number: 0, Flash bias: 0.00 EV, Subject Distance: 65535, File number: 000: 0000, Image Type: IMG:EOS 300D DIGITAL JPEG, Firmware Version: Firmware Version 1.1.1, Camera Serial Number: 1570502381, Image Number: 2868699, Owner Name: MarjaN

Figure 4-3. EXIF metadata as stored by a digital camera in a photo.

Besides at the point of creation, automatic metadata can also be added at a later stage. It may be based on mechanisms that first recognize some specific features from the content, and then translate them to metadata symbols. It is feasible to develop systems that can accurately count the number of people in a photo, but identifying whether one of them is Michael Jackson is more difficult.

The challenge of such automatically entered metadata is that it (currently) captures characteristics that are technical and therefore irrelevant to most users. Of course it may be possible to derive further information based on automatic metadata; a simple example is that date of creation implies the season, or that the use of a flash can imply dark conditions. However, deriving such metadata is still largely experimental, and requires advances in pattern recognition, context awareness, and artificial intelligence in general. We will discuss the potential of deriving metadata in section 4.4.

As with any automation, the major pitfall of using metadata to automate operations is that if the criteria for the operations are not clearly visible, the resulting view may be difficult for the user to comprehend – especially if the criteria involve multiple attributes that are not all visible. As a result, the users become frustrated as they cannot locate a certain item, or the grouping of the items is not obvious.

The history of an object can provide new metadata. It can be useful to track the user's interaction with the object and attach that information incrementally based on the various moments in the object's lifecycle (Figure 4-4). In this way the system can identify what objects are used often, which in turn indicates their value and importance to the user. In addition to time and date, the interaction metadata attributes can describe the action (such as open, edit, send) and parameters (such as to which telephone number the image has been sent). Note that such metadata creation takes place after the content object itself has been created.

There are networked services that provide metadata related to the content object automatically. Today, this is routinely done in the music management software for MP3 players. The metadata in the files typically comes from Internet services, such as freeDB.[21] CD ripping software acquires the metadata from the services and inserts it into the ID3 attributes of the MP3 files. As soon as one user has filled in the metadata for a new CD, it becomes available for other users of the same CD.

---

[21] http://www.freedb.org/

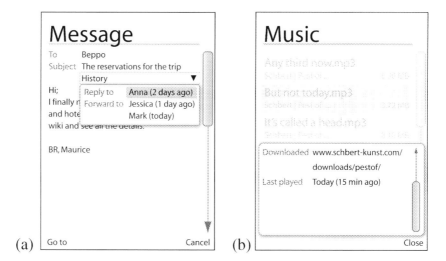

Figure 4-4. (a) Keeping track of messaging traces, here showing where a message has been forwarded, (b) keeping track of where an MP3 song was downloaded from.

The case of MP3 files also illustrates why mobile content management is becoming increasingly important. Consider the amount of content that users can carry. The maximum storage capacity of a portable MP3 player using flash memory increased from 1 GB to 16 GB between 2004 and 2007. As this increase shows no signs of stopping, this means that by the year 2010, the user may have to manage over 60 000 songs stored in the device.[22] Fortunately music is organized in a logical hierarchy by artist–album–track, and the mentioned services readily provide us with that information. The hierarchy can be automatically (re)generated with the help of metadata in the MP3 player. Forming a hierarchy and presenting the content as a tree helps the user to browse and locate a content object (Figure 4-5).

The issues related to automatic metadata creation vary. For instance, one should carefully consider the implications of false metadata, that is, the consequences of a situation in which the automatically generated metadata is incorrect. In many personal content-related cases, the result is not severe. For example, the number of people in an image might just exclude the photo from a scenario which it belongs to. Or, if you invest 99 cents into a song based on some misleading metadata and then realize that the content is not exactly what you expected, the economical loss is regrettable but not lethal.

---

[22] We applied the same formula that Apple uses for their iPod: one GB holds 240 songs, which are 4 minutes long and encoded with 128 kbit/s AAC coding.

Figure 4-5. Song metadata (content type, artist name, and album name) applied to maintain a folder hierarchy).

4.3   Metadata Maintenance

As you may have guessed, the metadata is not always static, but it changes as the content is used. For instance, some media players count how many times a track has been played, and based on this so-called play count then change the rating of the song. This makes sense, because it is unlikely that the users would torture themselves with bad music repeatedly. Naturally, such auto-rating should not override any manual ratings.

TO WINAMP OR NOT TO WINAMP?

Eddie finally came in with his portable disk and uploaded his 85GB collection of MP3 files into her computer. Okay, it might be illegal, but who cares – everybody's doing it. Eddie looked at her importing the MP3 files into iTunes, Angie's favourite player, and suggested that she switch to Winamp, which is just a way cooler piece of software. Angie wasn't sure. Eddie explained that he had all those carefully crafted playlists for Winamp, mixing the files he just gave to Angie into really spacey all-night sessions, and Angie should really try it out.

   Angie did install Winamp with Eddie and has opened it a few times since then. She still doesn't know . . . well, in fact she does. That software just doesn't feel right to her. What's worse, Angie has spent years in rating her song files with

---

iTunes, and quite rarely hears any bad pieces anymore. But if she goes for Winamp, all those ratings seem to be lost.

Wouldn't you know it, just after installing Winamp it started – most embarrassingly – playing her forgotten Backstreet Boys disks. Eddie said nothing but did raise his eyebrows disapprovingly. Well, now those files are gone, but shouldn't there be a way to export ratings from iTunes to Winamp?

After a few days of fiddling with Winamp, Angie gives up. She will stick with iTunes. She's got so much content already in that library that she could not be bothered to switch. Maybe Eddie could get an iTunes converter for his precious playlists?

In addition to mere insertions, metadata maintenance will also include deletion, in case of false or outdated metadata, or due to privacy reasons, and refinement such as analyzing metadata to make further assumptions.

When metadata is transformed between different systems, or when files are converted from one format to another, the metadata may have to be translated too. Often different systems have different naming conventions for similar metadata attributes, or the interpretation of the attributes differ between the systems.

Metadata is more than just attributes. An important source of information comes from inspecting content objects and finding similarities and other relations between them. Next, we will discuss the importance of relations as a form of metadata.

4.4   Relations Give Meaning

Object-level relations are essential building blocks for exploiting metadata for the user's benefit. Metadata is often seen as attributes of content objects, but there is also valuable information in the relations between the objects. For example, a song *is part of* an album, a book *refers to* another book, and an actor *is starring in* a film. Once we know the relations, we can trace the object networks that are formed by these links. Such networks will not only reveal items that are somehow related to the object at hand, but can be used in a number of ways to gain new information on the present object. Relations then give context and meaning to objects. Relations also enable *associative browsing*, helping the user to find objects that are logically connected in some way.

In some simple cases, relations allow constructing playlists, slideshows, or other one-dimensional collections. In more advanced cases, through the analysis of multiple content items, they can be exploited to construct automatic or semi-automatic stories about past events and experiences. In case references to other people are encountered, the analysis can also be expanded to these persons' personal content, if they so allow. Such rich networks that bind together social relationships and the related content objects truly enhance our abilities to enjoy and re-live past events.

The attributes and relations of metadata together form a language for talking about data. In computer science, such languages have been pioneered in the subfield of artificial intelligence (AI), known as *knowledge representation*. That area of research seeks to define ways of expressing data (or metadata) in such a way that certain computational problems become easier to solve. For instance, to solve a navigational problem on a map with multiple cities and roads connecting them, a useful data representation is a graph with cities as nodes and roads as arcs. Given the right knowledge representation, many algorithms become quick and powerful. Much of the research in AI continues to concentrate on finding suitable representations of the wonderfully complex world we live in.

We feel it is useful to regard metadata formats as knowledge representation languages. While not really aimed at generic AI problems, such languages can be indispensable for the more specific tasks that are needed in manipulating content. Particularly, as we are trying to take the content manipulation from mere computation towards a more user-oriented semantic interaction, it will become necessary to capture some of the semantic information of the content objects. Relations are an essential part of that information, as are the attributes.

Most metadata representations and formats have remained focused on attributes. Where relations are retained, they are frequently converted to text strings (or, at best, URLs) that are usually targeted for human consumption. That works if relational metadata can be omitted from automated analyses. We choose instead to encode both attributes and relations with machine-readable symbolic identifiers for later automated use. Chapter 5 describes our design approach in detail.

4.4.1 People as First-Class Metadata

Ideally, people should be treated as first-class metadata items (Figure 4-6). However, existing metadata formats normally reference people as text strings.[23] This works for simply displaying names, but falls short

---

[23] There are good reasons for this. It is hard enough to deal with swapping first names and last names, translating Chinese names to western representations, or track ladies' surnames through six marriages. Plain (Unicode) text is the least common denominator for naming people. The technical solutions for identity representation are still in the works, hopefully surfacing from the Internet community in the next few years.

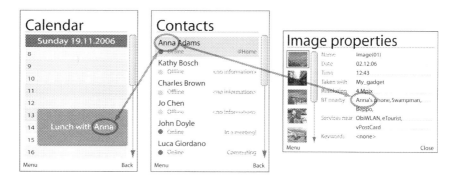

Figure 4-6. People should be first-class metadata items; all metadata fields should be interlinked.

for automated functions. For instance, since we store names of people in our contact lists on phones, PCs, and PDAs, we should be able to "click" them as easily as we click a file or a menu item. This would then allow accessing all the content that relates to these people.

At the moment, some software suites (such as Apple iLife or Microsoft Office) do treat people as objects that can be used as a basis for interaction. This treatment is not uniform, however. Most mobile phone UIs allow data access through contact lists, but fall short in some functions, such as linking content objects in arbitrary formats to contacts. We are promoting this view of people as first-class metadata citizens to be used throughout device and application UIs, database structures, logs, and wherever data is being manipulated. We expand this view in our metadata architecture described in Chapter 5.

4.4.2 Derived Metadata

All information describing a piece of content is metadata. However, as we wish to separate a special kind of metadata from other metadata, we call this metadata subset *derived metadata*. This term denotes metadata that not only talks about other data, but has been derived from other data through automated means. The relation of metadata and derived metadata could be described as:

- Metadata is *direct* data description: image resolution, bit rate of audio file, or a thumbnail image are typical instances of direct metadata.

- Derived metadata is *indirect* metadata: it may describe the links (associations) between objects, or contain some information calculated from direct metadata. For instance, the collection that an image belongs to is derived metadata, as is the listening history of an audio file.

Derived metadata seeks to raise the level of abstraction of the data in such a way that some of its aspects are closer to human thinking, as opposed to computer data structures. Today, such abstraction routinely takes place in pattern recognition, image analysis, speech analysis, statistical analysis, or any processing that translates raw data into meaningful indicators. As abstraction methods improve, we expect more analysis functions that can take personal content files as input and produce nearly human-level abstractions of those files as output. We call such functions *harvesters* and discuss them further in Chapter 5.

4.5   How does Metadata Benefit the User?

From the perspective of a mobile device user, the main advantage of metadata is that it makes personal content easier to manage and thus the device easier to use. The metadata facilitates a paradigm shift[24] from managing content instead of files and it has emerged because of many reasons (recall the discussion about file management versus content management in section 3.1). First, the number of files has skyrocketed and the files are often distributed in several devices. Second, because the content is created and edited by many people, mere file attributes (such as filename, date, size, and access rights) cannot describe the subtle differences between the content objects without understanding the information that the objects contain. For example, try telling the differences between the successive two images with names like IMG_2745.JPG and IMG_2746.JPG, without looking at the contents.

There are several uses of metadata. For instance, it is an important enabler for searches, especially when searching for media objects that do not contain any text. Furthermore, it can automate various tasks related to organizing, summarizing, and viewing of the content. Next we introduce some of those operations that profit of metadata. Our treatment of these operations builds on the classification by El-Sherbini (2000).

---

[24] A "paradigm shift" refers to a fundamental change in thinking. We believe that focusing on content instead of files is a fundamental change indeed.

4.5.1   Tracing and Recall

The simplest way to use metadata is simply to store additional information which describes the content and actions performed on the object. This is particularly beneficial for content formats (such as images and video), which do not contain machine-readable textual data. The additional information is also used for providing information for data catalogues.

Furthermore, metadata can describe the creator of the content and the history of data, such as the original source of the data and any subsequent transformations that can aid in tracing the content lifecycle. The benefit of history information is that it assists the user in recalling the context (for example, people, event, and location) as well as helping content management systems to trace the changes between multiple copies of the same content.

4.5.2   Searching

Searching is perhaps one of the most typical examples of operations that benefit from metadata. When metadata is used for searching, it aims at decreasing the amount of work required for formulating the query and enhancing discovering by rapidly providing only relevant and useful results. Essentially, this is due to the fact that it is easier for a computer to search for symbols than recognize objects and features from raw multimedia content. Looking up "Person: John Smith" from an indexed database is thousands of times faster than scanning images for faces and matching those to Smith's facial features.

Some search mechanisms that metadata facilitates include:

- *Search-by-example* allows users to search without defining explicit keywords. Instead a user can select an object that provides an example of content that the user wishes to discover. For example, this allows finding a video of football games by showing a clip of a Tottenham game. This can also be referred to as relevance feedback.

- Creating *personalized searches* that are based on the user's prior behaviour or activities.

- *Ad hoc searches* for content that is not indexed. For instance, while walking on the street, the user could ask the device to find the latest message from Eddie that contains Angie's new telephone number.

- *Collaborative searching* is using other users' previous searches on a same or similar topic. It can be based on, for instance, examining the other users' interaction with the search results that they have received after making a similar query (Lehikoinen et al. 2006).

- *Proactive searches* based on assumptions of future activity. For example, the system could explore the events in the user's calendar, try to identify what content is related to the events, and download it to the device or – at least – provide a shortcut to the object.

Searching is discussed further in Chapters 5, and from the human perspective in Chapter 6.

4.5.3 Organizing: Sorting, Grouping, and Filtering

Metadata allows people to organize their content in a number of different ways. Which way is most useful depends on the situation and personal preferences. For instance, although a date is often a good way to organize images, people might prefer to look for content based on events, such as "Holiday in Italy" or "Dad's birthday". Users should to be able to change the way content is organized by using different criteria, such as creation time or content type. Organizing includes the following operations:

- *Sorting* is ordering objects based on values of some attribute, known as a sort key. There may be a single sort key (attribute), or more complex sort ordering may be done by combining several keys. The objects can be arranged in ascending or descending order, depending on the values of the sort key and its type (numerical, alphabetical, time).

- *Filtering* is determining whether some objects fulfil a certain criteria or not. This helps the user to focus on some objects by hiding other objects that do not fulfil the criteria. Filtering is typically based on checking whether the object's attribute has a certain value or if the value is within some range.

- *Grouping* is about finding some common characteristics (or values of attributes) from a set of objects and arranging these objects together. For example, many current audio players create automatic playlists based on the user's rating of the songs. The content type affects grouping the objects.

Automatic organization is easier for some content types than for others – for instance, personal images and videos versus commercial music. Images are often related to some special occasion (such as birthday or holiday trip) or place (such as home or a theme park), which are difficult for the device to know even though they may be obvious to the user. The temporal dimension is perhaps the easiest to take into account in organizing an object; just study the deviation of capture times of the images. Some other aspects would benefit from using location information or discovering other devices nearby.

4.5.4   Automatic Summarizing

Automatic summarization of content is related to grouping, but its aim is different as it goes one step further. As grouping is targeted at creating collections of objects that share similar characteristics, summarization is about finding, for instance, the most representative or novel object from the collection, or condensing a set of objects into a single summary object.

Consider image grouping. Automatic summarization selects an image that presents the group in the best way. The result can provide unexpected and fresh views to existing content – or sometimes fail completely, as is normal for automatics that operate outside the expected conditions.

One use for auto-summarization is storytelling: sharing personal experiences by generating an instant story based on a large set of objects. For instance, instead of going through the gallery of your holiday trip and wondering what to show, the system can generate a story based on the time, the travel route, and the captured content. The result may not be perfect but still sufficient to keep your politely interested dinner guests from falling asleep in the middle of your slide show.

4.5.5   Enhancing Privacy and Security

As mobile devices are becoming more vulnerable to different kinds of threats (such as viruses and malicious software), finding new ways to enhance security is increasingly important. Metadata can be used towards this aim by preventing the execution of some operations automatically, or blocking content that does not contain a certain combination of metadata attributes and values. Furthermore, the system could use a third party to check the metadata and determine whether the content can be trusted or not. Of course, also the metadata is itself content that needs to be protected.

User privacy can be enhanced by excluding private data from sharing or publishing on the basis of metadata. For example, a privacy rule might dictate that if the metadata contains the name of the company where the user works, it cannot be sent to e-mail addresses that do not belong to the same organization.[25] Further metadata may contain information, such as legal conditions and copyrights, which affect how to use that data.

4.5.6 Constructing Views

The metadata plays an important role in constructing UI views to objects. For instance, it can help in selecting what objects are to be displayed, how they are organized, and what information per object is presented.

In addition, metadata can be used for generating the UI layout. For example, Context Comic (Aaltonen et al. 2005) diary application uses metadata for selecting the personal content objects that belong to the diary and forming groups of objects based on diary events. For each event, the event's importance is derived from metadata attributes of its objects (such as number of times that the object has been opened or sent) and the same information is used for selecting the object that visualizes the event. The application further scales images and places them on a grid on the basis of the event's importance.

4.5.7 Better Recommendations

Metadata could also fine-tune the recommendation systems by taking into account context and comments by others people who have similar preferences. For instance, in a movie recommendation service, you might favour ratings from people who have seen the same movies you liked. Metadata could also link the recommendations to the individuals who contributed them so that their digital reputation could affect the ranking.

4.5.8 Reusing / Remixing / Reconstructing

Digital content readily lends itself to reuse, if the copyrights allow. Thanks to this, more and more of contemporary content are collages. Consider a typical music video on Music TV. Often the music track is created using dozens of sound samples obtained from commercial sound libraries, and the same may apply to the video files and visual effects. The metadata in these files may retain references to the original

---

[25] Which may not be quite what you want if, for instance, you are sending some promotion material to your customers. Unfortunately, simple constraints like this will often fail with the complex situations found in everyday life.

sources, which can then be credited in the final result for the purposes of tracing, charging, or rights management. The individual artists and the production crew may also be referenced in the metadata.

Usually production data will not be as interesting to the general public, but some devotees will spend much time in analyzing and dissecting their favourite artists' works, contributing new metadata in the process. For instance, the fans of *Star Wars* have built numerous fan sites where the movies are being recreated.[26] The dialogue, the plot, and the characters are analyzed in breathtaking detail, and sometimes that analysis is used to produce variations or parodies of the original *Star Wars* films. Such fan fiction is a form of participatory culture (Jenkins 1992).

YOU ONLY LIVE N TIMES

Cathy is intrigued by the hit song playing in her cellphone radio. The string theme reminds her of something . . . must be an old song she knows, but the name escapes her. As she hears the same song again for the third time the same day, she decides to check the channel's playlist. She browses the "Recently played" history in her phone. Aha, this must be it . . . "Millennium" by Robbie Williams. Ok then. But what about that string theme? Her friend Zach says that Mr. Williams likes to feature samples of old hit songs, so this prominent theme must be a sample. But which song would it be?

After some days of wondering, she meets Zach over coffee and tells him about her problem. Zach shows how she can check the channel's Web site for that playlist, then click on the song name and get it's metadata. Sure enough, the data shows that the song contains a sample of the theme to James Bond movie *You only live twice*. That's it! Not a great flick, but the song is just unbelievably beautiful.

Zach explains that if she got herself a smart phone with a decent browser, she could access that metadata immediately upon hearing the song. Cathy could not be bothered to get a new phone, but she still decides to download the sample from the channel's site as a new ringtone for her current phone. It's such a cool sample – not because of Mr. Williams, but for the song. The singer is quite amazing. It must have been Shirley Bassey at her best. Maybe Cathy could dig up more info on her?

4.5.9 Smoother Transition Between Applications

Metadata allows the same content to be more easily used in different applications. For instance, the popular GarageBand application by Apple allows users to create podcasts in a streamlined manner. The

---

[26] See, for instance, http://www.theforce.net/

content created with other iLife suite applications (photos, videos, songs) is all easily accessible in the Garageband browser for incorporation into the podcast under editing. Here, Apple has raised the abstraction level from file management into content management.

The metadata that the iLife suite uses is not directly visible, as it is used by the applications, not the users. This software suite was written by a single manufacturer, so the metadata formats were easier to agree upon. In cases where multiple companies are involved, agreements on the formats can become excruciatingly difficult.

4.6   Existing Approaches

In the following we review some of the existing metadata standards. We have chosen these, either because they are popular with consumer devices or because they illustrate technical aspects relevant to our purpose. Many other formats exist, so this treatment should not be considered as an exhaustive list. For a review on additional current metadata standards, refer to Smith and Schirling (2006).

```
COMMENT: MOBILE MEDIA AND METADATA

The increasing number of media creation devices is adding
to the organization and media management problem for per-
sonal media. Content describing metadata is seen as the
most prominent answer to this problem. However, the
current media content metadata standards, such as EXIF,
Dublin Core, MPEG-7, and IPTC, have shortcomings when it
comes to the domain of personal media. This is because
the standards were designed for public and professional
uses of media (e.g., public archives, commercial uses,
professional media creation). They were not designed with
personal media or user-generated media in mind.

In personal media some fundamental principles about public
and professional media and metadata do not hold true:
Personal media and metadata are mostly private, not
public. They are highly semantic, because the meanings
associated to them are personal, and this means that
automatic metadata generation is more difficult. Personal
uses are different from public uses, for example, personal
media is often shared in a private context within a social
network. Lastly, metadata is a non-trivial concept for
the average consumer to learn. None of the currently used
media content metadata standards were designed with these
principles in mind.
```

One of the paths not fully explored in personal media
metadata is to leverage social activity as a metadata
resource: the sharing, discussions, and popularity information already available in many services. Social activity
around personal media is important to the end-users and
formalizing it into metadata would enable new kinds of
approaches for solving the personal media management
problem.

Dr. Risto Sarvas
Researcher
Helsinki Institute for Information Technology HIIT,
Finland

(*Dr. Risto Sarvas' work has focused on designing user-centric metadata for digital snapshot photography. His current research goal is to bridge the gap between photography studies in technical, cultural, and social sciences.*)

4.6.1 MARC

Data about data has existed practically for as long as data has existed. The history of metadata can be tracked back to the days of the first libraries, where librarians catalogued the books in the collection, making it significantly easier to locate a desired book. Obviously, they were not thinking in terms of metadata, but rather indexes or catalogues.

Not surprisingly, then, one of the first metadata formats is MARC (Machine Readable Catalogue),[27] a format devised to transfer catalogue records between library systems and based on age-old cataloguing cards used in libraries (Dovey 2000). Several variations of MARC exist, such as UKMARC (developed and promoted by British Library), MARC21 (maintained by the Library of Congress in the US), and UNIMARC.

Typically, one MARC record consists of a number of fields, labelled from 0 to 999. Each field has a number of indicators that define its semantics, and a number of subfields. For instance, field 245 indicates the title, and the subfields indicate any additional information related to the title, such as additional title, statement of responsibility, and so forth (Dovey 2000).

MARC records are stored in binary format, even though they can be also rendered textually, which allows them to be edited directly

---

[27] Please see the following Web link for detailed information on MARC: http://www.loc.gov/marc

with any text editor. In addition to the transfer format itself, MARC also requires cataloguing rules that define the semantics of various fields.

MARC was defined in 1960s, and is well-established with thoroughly thought-out rules and semantics that have evolved over centuries along with library cataloguing. Furthermore, since the format is extensible, and the semantics are not tied with the syntax, it is easy to expand when needed, and is usable across several different types of multimedia objects, not just books.

From the viewpoint of personal content, MARC feels like overkill – or not enough. Most people are not interested in classifying their content with scholarly accuracy. Instead, they want to focus on the enjoyment, experiences, and emotion. While MARC pioneered many of the concepts that newer formats use, it does not seem to raise enough interest for the mobile content domain. Or, it may just be that librarians do not speak with people on the Internet and computing domain – and vice versa.

4.6.2 Dublin Core Metadata Initiative

One of the most important efforts for creating a standard for general resource description is Dublin Core Metadata Initiative (DCMI).[28] As stated in its home page:

> The Dublin Core Metadata Initiative (DCMI) is an organization dedicated to promoting the widespread adoption of interoperable metadata standards and developing specialized metadata vocabularies for describing resources that enable more intelligent information discovery systems.

The mission of DCMI is to provide simple standards to facilitate finding, sharing, and management of information by the means of developing and maintaining international standards for describing resources, supporting a worldwide community of users and developers, and promoting widespread use of Dublin Core solutions.[29] This is an ambitious goal, considering all potential information sources and backgrounds of the content creators, users, and distributors with a variety of needs and expectations.

DCMI is not a metadata standard as such, but instead aims to maintain a core set of *metadata terms*.[30] The metadata terms, then, include

---

[28] http://dublincore.org/
[29] General description of Dublin Core: http://dublincore.org/about/
[30] DCMI Terms: http://dublincore.org/documents/dcmi-terms/

elements, *element refinements*, *encoding schemes* (qualifiers), and controlled *vocabulary terms* (the DCMI type vocabulary). For each term, some metadata (or meta-metadata) is included: Universal Resource Identifier (URI), label name, its definition in human-readable form, an optional comment, type (element, element refinement, encoding-scheme, vocabulary term), status, and date issued.

The elements include labels such as contributor, coverage, creator, date, description, format, identifier, and language. Element refinements include labels such as access rights, audience, dateCopyrighted, and relations such as isPartOf and isReplacedBy. Encoding schemes include labels such as IMT (Internet Media Type), ISO3166 (country codes), MESH (medical subject headings), DCMI Point (a point in space), and URI. Finally, vocabulary terms include labels such as Collection, Dataset, Image, InteractiveResource, MovingImage, Service, Software, and Text.

DCMI has not gained a strong foothold in media industry,[31] even though several global projects are applying it. It seems that the need for cross-media metadata formats, or even formats that span across virtually any media type, are not considered vital by the major players in the field of digital media. Obviously an all-encompassing content metadata format is not practical, as the field of digital content is developing rapidly. New formats and conventions come and go on a monthly basis. However, it appears DCMI is sufficiently powerful and relevant that almost any content must inevitably include its definitions, either directly or redesigned.

In the recent years, some professional content producers have started to realize the potential of cross-media metadata and relations between content objects in different formats. Adobe XMP is an example of a cross-media metadata platform that has been adopted by several important players in the industry.

4.6.3 XMP

Adobe's XMP (Extensible Metadata Platform)[32] allows embedding metadata into content files and provides access to it. Furthermore, it defines standard cross-media metadata schemas, such as Basic Schema (influenced by Dublin Core), which can be used with any type of media. XMP also includes Rights Management and Media Management schemas. There are currently several standard bodies working on

---

[31] How widespread is DCMI: http://askdcmi.askvrd.org/default.aspx?id= 11112&cat=1739

[32] http://www.adobe.com/products/xmp/

initiatives based on XMP,[33] such as Creative Commons, IPTC, and W3C to name but few (Rosenblatt 2002).

XMP is not tied to single content type, and is built on open W3C standards: it is an implementation of RDF (Section 5.7.3). Furthermore, XMP is extendable, also by third parties. It is especially targeted at creative professionals. XMP consists of four building blocks:

1. XMP Framework: RDF Framework for expressing metadata from multiple schemas;

2. XMP schemas: Schemas used to describe properties, contained in namespaces;

3. XMP Packet Technology: Method for embedding XML fragments in binary streams;

4. XMP SDK: Support for third party interface and extensions to XMP.

Of the existing metadata content standards to date, XMP appears to have the most flexible design, in terms of extensibility and application interfacing, while tying to the (potentially huge) RDF-based information sources in the Semantic Web.

XMP has also subsumed an earlier *de facto* standard, "IPTC Headers" that Adobe adopted in their Photoshop image manipulation software, particularly used in professional digital photography. The headers embed metadata into image files in JPEG, TIFF, and Photoshop formats. Other vendors have also adopted the use of these metadata fields in their image editors. The key use of the metadata structures is multimedia news exchange between professional news agencies and publishers, which is reflected in the choice of the metadata attributes.

The IPTC headers are based on an Information Interchange Model (IIM)[34] that was designed for universal data transfer of graphical, textual, or other multimedia information. It is structured as an envelope around a data object, or collection of objects. This general format allows encapsulation of any metadata with any objects; however, the main use to date has been in the professional news domain.

Besides generic metadata initiatives, there are several metadata formats that are more or less tied to specific content types. The best-known for general public are undoubtedly MPEG-1 Layer 3 audio media type with its embedded ID3v2 tag information (such as song

---

[33] http://www.adobe.com/products/xmp/standards.html
[34] http://www.iptc.org/IIM/

name, artist, album, and genre), and Exchangeable Image File Format (EXIF) metadata for still images. Since music and images also are the most common content types GEMSed by the general public, we will now discuss these metadata formats in more detail.

4.6.4 ID3v2

Augmenting music files with metadata is often referred to as tagging, and the most important and best-known format is ID3; the latest version towards the end of 2006 being ID3v2.4.0.[35] The term was originally coined by Eric Kemp in 1996, where the acronym comes from the words "IDentify an MP3". Other widely used audio tagging formats are those used by Microsoft's Windows Media Audio (WMA) files, and AAC (Advanced Audio Codec) files. AAC files are most often encountered with Apple's online store iTunes,[36] where a rights-protected version of the codec is used.

ID3v2 metadata is stored physically in the same files with the actual content, which ensures that metadata travels with the content with no extra effort when the files are copied or moved. This is in many cases considered a plus, as information related to the song itself is not lost. However, since the tags may also contain personal or device-dependent information, such as ratings and play counts, it is not always appropriate to transfer all metadata with the content. For instance, when transferring a song from a desktop computer to a portable music player, it may or may not be desirable that the play count indicator is retained. As a result, some rules are required to keep the record straight.

ID3v2 is a container format, in practice a block of data added before the binary audio data (Figure 4-7). Each ID3v2 tag holds one or more smaller chunks of information, called frames. These frames can contain any kind of information and data, such as title, album, performer, Web site, lyrics, equalizer presets, and images. The block scheme above is an example of how the layout of a typical ID3v2-tagged audio file looks.

ID3v2 metadata format is extensible, which allows new metadata tags to be included when required. Each tag is encoded in frames with predefined identifiers, and new ones can be defined as needed.

ID3 tags have arguably brought metadata fields to the mainstream public. Practically all music management software makes extensive use of these tags, which have now reached an industry standard status.

---

[35] http://www.id3.org/
[36] http://www.apple.com/iTunes

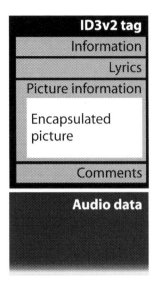

Figure 4-7. The layout of an MP3 file with ID3 tags.

4.6.5 Acidized Wav

Contemporary music production is based on loops, particularly for the great majority of songs in the western Top50 charts. Pop songs typically consist of individual audio samples that are repeated and recombined using a variety of methods, including sample playback and sample re-synthesis. Traditionally this was done in expensive commercial music studios, but in the last decade hundreds of more affordable software tools have appeared that allow hobbyists, semi-pros, and amateur musicians to craft technically advanced pieces of music in their bed-rooms. Most professional musicians now have smaller so-called "project studios" for creating music on their own. This has resulted in an esti-mated 350 000 project studios worldwide.[37] Many of these studios use existing sound samples as their source material. Such samples are avail-able from a variety of sources, including sample collections on CDs and downloads from the Internet.

The ACID file format for audio sample files quickly became popular with both professionals and amateur musicians when it was established in the late 1990s. Designed by Sonic Foundry for their ACID audio manipulation software, the format embeds extra information about the

---

[37] Source: Anthony Grimani, Performance Media Industries (PMI) Ltd, Fairfax, CA, USA, in AES 121 conference in San Francisco, October 2006 (http://www.aes.org/121)

audio within Wav files (the *de facto* audio file format for Microsoft Windows). This information segments the audio data within the sample into smaller chunks, so that each chunk can be looped semi-independently of the other chunks within the file. As a result, the tempo and the pitch of the audio sample can be varied freely, allowing for greater possibilities for sample manipulation.

Before ACID, tempo and pitch could be manipulated independently only with very expensive studio equipment. The development of this format, together with the audio stretching software for it, made real-time audio stretching available to the masses. This functionality has later been incorporated into numerous music software packages, such as Apple's Garageband. This, in turn, has contributed greatly to the art of real-time mixing and to the currently popular surge of "mashups", where two or more songs are combined in unorthodox ways.

While the ACID format is not exactly a consumer format – in fact just an extension of a consumer format – we can say it provoked a fundamental change in the way musicians dealt with sound samples, and consequently brought the technology within the reach of consumers. A lesson to learn here is that sometimes minor additions of metadata into existing formats can have major effects.

4.6.6 DCF and EXIF

As discussed in Chapter 1, the increase of digital camera sales during the past few years is nothing short of astonishing. As a result, our mass media devices are filled with digital photos. Since most current digital cameras include metadata in Exchangeable Image File Format (EXIF) formats, it is not surprising that EXIF is the dominant image metadata format at the moment. The current version (2006) is EXIF2.2.

The EXIF format is part of the DCF (Design Rule for Camera File Systems) standard created by the Japan Electronics and Information Technology Industries Association (JEITA)[38] to encourage interoperability between imaging devices. DCF is a *file system standard* designed to ensure compatibility for images using recording media (for instance, Flash memory cards) between devices such as digital cameras and printers. The standard defines file naming conventions and folder structures. DCF-compatible devices allow image files to be easily exchanged, even between different series of products or the products of different manufacturers.

---

[38] http://www.jeita.or.jp/english/index.htm

EXIF, then, is a standard for defining how to store image data and camera information in files, and applies to the actual data itself. EXIF information is embedded in the files along with the actual content. As can be seen from Figure 4-3, EXIF deals mostly with camera-based technical information, such as the use of flash, aperture size, and zoom level. A different approach was adopted by IIM/IPTC headers, which we discussed above. These image formats all largely lack some of the potential items of interest to the users, such as automatically entered context information.

4.6.7 Quicktime

The Quicktime multimedia architecture[39] developed by Apple comprises a rich set of conventions, file formats, APIs, and tools for creating, viewing, and editing multimedia content on personal computers. It has been mostly used for movies (famously for film trailers and music videos), but allows synchronization, compositing, layering, and control of multiple kinds of media, including images, audio, text, sprites, live streams, synthesized music, VR, and others.

While the Quicktime format is not specifically designed to convey arbitrary metadata about media files, its structure allows embedding any kind of metadata. A Quicktime "movie" consists of multiple tracks, which are often time-synchronized, each of which can contain multiple instances of a single media type with different characteristics. The format stores various kinds of metadata with each media segment, including dimensions, frame/sample rates, timescales, codecs, non-destructive edit lists, etc.. In effect, Quicktime playback consists of a real-time multitrack, multiformat, multimedia, re-synthesis process.

Quicktime files are organized into hierarchical trees of "atoms", that in turn refer to other atoms, until the leafs on the tree contain the elementary information, such as movie frames. Metadata is then stored within such atoms, as with any other content. The Quicktime APIs allow higher-level manipulation of the known metadata types. The file organization and object nesting relies on keeping track of object sizes as they are mapped into linear files.

Quicktime supports nonlinear datatypes, such as "sprites", which can be useful in allowing user control of playback, or to refer to external information in specific places in the movie ("Click the actress for her biography").

The Quicktime architecture achieves many desirable goals of a metadata architecture: it is extensible, transparent, efficient, and it

---

[39] http://developer.apple.com/quicktime/

works over several kinds of media and platforms. However, it has mostly been targeted for use by professional multimedia creators. It has less relevance for user-level metadata in consumer created content – although the architecture does not exclude that use.

4.6.8 MPEG-7

MPEG-7 is part of the MPEG family of standards[40] defined by the Motion Picture Experts Group within the ISO standardising organization. This standardizing body is responsible for defining many of the formats described elsewhere in this book, such as MP3 (more precisely: MPEG-1 Audio Layer 3). The standards from MPEG include MPEG-1 and MPEG-2 for audio and video compression, MPEG-4 for more complex audio and video documents with object-based composition, 3D rendering, interactivity, and other features, and MPEG-21 for handling digital multimedia items with integral copyright management throughout content production, distribution, and consumption chains.

MPEG-7 is closely related to the domain of this book. It defines ways of expressing multimedia content metadata (not the content itself, as with MPEG 1, 2, and 4). This rich, ambitious framework defines a range of constructs such as descriptors, description schemes, definition languages, etc. – essentially all components of a complete metadata system (Manjunath et al. 2002). Moreover, the design of MPEG-7 seeks to ensure that the resulting descriptions are extensible, layered, embeddable, streamable, distributable, semantically rich, and relevant to a wide range of applications.

All these admirable goals have led the standard to appear complex by its design, probably reflecting the complexity of the domain. No other multimedia standard succeeds in grasping as many conceptually demanding issues within one framework. This may also be a key weakness of the standard, as it appears to be sufficiently rich to describe almost any existing and upcoming piece of multimedia.

MPEG-7 tackles the apparent complexity by proposing profiles for specific applications, such as bibliographical references. Nevertheless, commercial support for the standard currently seems somewhat modest,[41] and development of the standard at the moment has ceased.

---

[40] http://www.chiariglione.org/mpeg/
[41] http://www.chiariglione.org/mpeg/working_documents/mpeg-07/general/industry_usage.zip

4.6.9 RSS

The RSS metadata format (Really Simple Syndication, or Rich Site Summaries)[42] has become a *de facto* choice for describing the contents of frequently changing Web pages, such as blog entries. It has also been adopted for use in podcasts, news clips, image galleries, even weather forecasts, and any number of data sources available on the Internet.

RSS stores metadata that effectively allows quick summarizing of media items. It defines a number of fields encoded in XML, such as titles, Web links, contacts, timestamps, textual summaries, etc. This allows client software, for instance a newsreader, to quickly check for new content on a page on behalf of the user.

An example of a RSS feed (feed://rss.slashdot.org/Slashdot/slashdot/to)

```
<div name="0/3/29" class="read"
    onmousedown="return articleClicked(this)"
    articlesortdate="186537244"
    articlesorttitle="acoustic levitation"
    articlesortsource="" sourceindex="">
    <div class=articlefooter></div>
        <a   class=articlehead   href="http://rss.slashdot.org/~r/Slashdot/
slashdot/to/~3/55605066/article.pl" target="_self">
            <div class="subject"
                title="Acoustic Levitation Works on Small Animals">
                Acoustic Levitation Works on Small Animals
            </div>
            <div class="date" title="Today, 03:54 ip.">
                Today, 03:54 ip.
            </div>
        </a>
    <div class=articlebody>
        <a
href="http://rss.slashdot.org/~a/Slashdot/slashdot/to?a=mDbmZD">
        <img
src="http://rss.slashdot.org/~a/Slashdot/slashdot/to?i=mDbmZD"border="0"/>
        </a>
        <img
src="http://rss.slashdot.org/~r/Slashdot/slashdot/to/~4/55605066"/>
        <a  class="articlelink"  href="http://rss.slashdot.org/~r/Slashdot/
slashdot/to/~3/55605066/article.pl"
                target="_self">Read more&hellip;</a>
    </div>
</div>
```

Since RSS defines both fields that are, in principle, readable by both humans and machines, it manages to capture the best of both worlds. It allows machines to quickly process mechanical data (i.e., timestamps) and lets humans concentrate on the actual content. It is therefore a good match to describe frequently changing news-style

---
[42] http://web.resource.org/rss/1.0/

content that can be easily summarized due to its hierarchical structure.

The metadata fields defined by RSS have already been used for a number of purposes beyond news summaries, for instance stock charts, book reviews, comics, price trackers, classified ads, and others. This relatively lightweight metadata standard has arguably allowed a surge in the machine-readable Web content – including personal content.

4.6.10 Summary

Unfortunately, the field of multimedia metadata standards is unclear and immature at best. The sheer number of different media formats alone makes it a real challenge, not to speak of the huge data volume and the number of practitioners in the field, and the number of fields.

Adoption to different proposals has been slow, and vendors provide their proprietary, application-specific solutions. It is evident that no single body can take care of all metadata standardization efforts: the domain is simply too wide (Smith and Schirling 2006). This calls for better co-operation between the standardization bodies and an adoption of a more modular approach.

4.7   The PCE Trinity: Mobility, Context, and Metadata

The discussions above reveal a significant fact: the existing metadata formats and standards do not consider consumer use by design. Some of them can be applied to personal use, yet inherently they are all targeted at content creation professionals.

What, then, would be needed in such a metadata standard? As we emphasized in section 4.4, the importance of people, that is, social relationships, is a key issue. Here, we will discuss another aspect, inherent in mobile use: *context-awareness* – the capacity of the devices to take into account the user's situation. These two aspects are mostly missing in existing metadata standards.

Mobility from the device perspective refers to something that is *capable of being moved*. This is a fundamental characteristic of a mobile device, since it distinguishes them from desktop computers. From the user's point of view, in section 2.2 we categorized mobile device use as continuous or nomadic, depending on if the device can be operated at the same time as the user is moving (for instance, by foot or driving a car). To emphasize, *truly mobile devices for continuous use are the ones that the user can use even while on the move.*

Typically continuous use consists of small operations that require a little time, whereas the user can focus on using desktop computer for longer periods. For instance, when the user is jogging and listening to music, probably they adjust the volume without needing to use their eyes, since attention has to be continuously paid to the environment.

In mobile use, the information related to the environment and its changes is often more significant than in desktop computing. Information related to the situation is called *context* and it may be defined as (Dey 2001):

> "*Context is any information that can be used to characterize the situation of an entity. An entity is a person, place, or object that is considered relevant to the interaction between a user and an application, including the user and applications themselves.*"

Most applications have little understanding of what the users are doing. Occasionally the mobile device may disturb the user's current task. For instance, the device may notify the user about a received text message while engaged in a conversation. In some other situations, using the device is not possible as the user is occupied with another task, such as driving. Thus, it would be beneficial for the user if the mobile devices had some awareness of their current context. Such systems are called *context-aware* (Dey 2001):

> "*A system is context aware, if it uses context to provide relevant information and / or services to the user, where relevancy depends on the user's task.*"

Context could be used to improve ease-of-use by decreasing the need for user input; reducing the amount of information that the user has to process; increasing the quality of presented information; adding context to information in order to support later information retrieval; and reducing the complexity of rules constituting the user's mental model of the system (Cheverst et al. 2001; Dey 2001).

There are many technologies for gathering context information, for instance, sensors embedded in the device, network services, and deriving the information from other information. The gathered information is usually processed on the background and if not used immediately, can be stored for later use.

4.7.1 File Context

Earlier we discussed how metadata may provide additional information about the content and enable various things that are beneficial to the

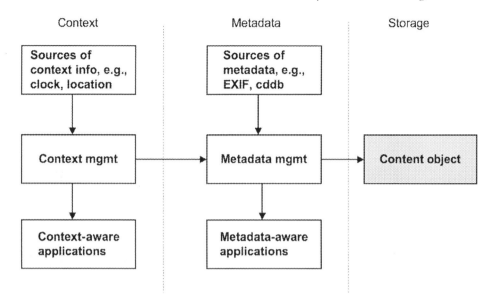

Figure 4-8. File context means that context information is stored in metadata fields of content files.

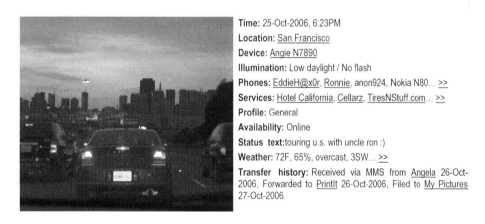

Figure 4-9. An example of file context. Compare to Figure 4.3.

user. However, we have just mentioned that context information characterizes the different aspects of an entity. So it makes sense to connect these two concepts and call the result *file context*. File context is explicitly stored metadata based on any context information, which was available at the time of capture or use, to the content object (Figure 4-8). As the manual creation of the metadata is a laborious task, file context eases the problem related to metadata entry as well as increasing the potential uses of the content.

The photo (Figure 4-9) was apparently taken by Angie on a holiday in California. All the fields in the metadata have been obtained automatically from her phone, surrounding devices, and network services available at the time and location where the image was shot. She had earlier entered the Availability and Status fields by hand for the purposes of the presence application, but the data have been reused in the camera application without additional user effort. Comparing this information to EXIF data stored by a digital still camera (Figure 4-3), one immediately notices a difference in abstraction level.

Further events in the life of the media object can also be recorded in the file context. Editing, viewing, messaging, and organizing the file can all leave not only interaction history, but also context history, which can be significant for later analysis. For instance, people often look for e-mail attachments, not by their names or their content, but by the name of the sender. It can be beneficial and interesting to recover the communication routes our important files have travelled.

In our photo example, the Transfer history metadata field shows how Angie's mother had received the file in a multimedia message, sent it for printing, and filed it for later viewing. This information can be used later to search for the photo when she cannot find the file but remembers that she received it from Angie.

4.7.2 Elements of Context

Context is a vague, hard-to-define concept (like metadata) that can be used to refer to almost arbitrary kinds of information. To highlight various elements of context, let us take a look at a more detailed categorization (ePerSpace 2004) and study what characteristics would be valuable for later use:

- *Environmental context*, which describes the entities around and the environment where the user and the device currently are. The information describes, for example, other devices, other users, brightness, temperature, and humidity.

- *Social context* defines the relationships between users nearby. It can contain information about friends, co-workers, relatives, and so forth.

- *Spatio-temporal* context means the location in the real world at a certain moment of time.

- *Task context* describes information related to on-going tasks, for example, based on the calendar information.

- *Personal context* refers to physiological (such as blood pressure and pulse) and mental context (for instance mood and stress) related to the user.

- *Terminal's context* relates to the device status and capabilities, for instance, characteristics of the terminal (screen size, weight, etc.), and status of power and memory.

- *Service context* describes what services are available and if the user is using some of them. This includes information, such as service name, state of the service (for instance, running or suspended), and cost of the service.

- *Access context* describes the status of the network connectivity.

In the previous list, the four topmost items are the most important for personal content experience. They are related to time, location, occasion, and other people, which typically characterize an event. However, the other items may be equally important for some applications.

As a general rule, what we consider the most crucial context categories are social context and location. These two primary context sources would probably suffice to deduce many of the essential aspects of the original situations.

4.7.3 Context is Essential for Communication

As an illustrative example of how context affects communication, consider a typical phone call. In the case of stationary landline phones, the caller could assume the location of the callee. When placing a call to an office number, we could assume that the other party was at work. Answering the home number, she would be in a home setting. Based on knowledge of the number we could already guess much of the other party's context.

The mobile phone has broken these fixed settings. People can place and answer calls anywhere. The potential contexts are now much more

Figure 4-10. Humans need to share context to communicate (photo: courtesy of Tero Hakala).

unpredictable. This could be one of the reasons why people tend to start their calls with dialogue that serves to establish context: "Guess where I am," "Are you in a car?", "Are you in a good place to talk?" It appears that in order to engage in dialogue, we need to gain some kind of understanding of the context on the other party.

Likewise, mobile devices will benefit from context awareness to be able to successfully interact with people in mobile situations. For instance, consider a mobile yellow page catalogue for finding nearby services, such as pizzerias. Without knowledge of the user's location, an electronic catalogue is not much better than a paper-based manual. However, upon knowing the user's location, the catalogue can sort the available services in terms of their proximity, giving fundamentally more useful information.

4.8   The Challenges: Universal Metadata, Extensibility, Abuse

Today, metadata is mostly local, in the sense that often the meaning gets lost when moved from one system to another. In contrast, universal metadata would be useful in more places. There are some metadata attributes that are used in next to all computing systems. For instance, filenames and creation dates are often retained when moving content files between systems. Still the dates may be interpreted differently – should the date be creation date, access date, or date of latest change? Perhaps all of them are needed. Filenames get truncated, version numbers are lost, creator information disappears, and so on. And this is just on a relatively well established concept – files. Things get worse when we are dealing with data that is less well understood, such as models of social networks.

NOT A GOOD MATCH

Cathy is again uploading scans of her paintings to their Web site. It really stinks that she has to use dedicated applications for each different task: scanning the paintings to her computer, browsing, editing, and sending the images. What's funny is that the scanner and the editor have different ideas of image orientation, so she always has to manually rotate each image before editing.

Also she has not been able to figure out how to set the image size properly. If she prints drafts from the editor, it comes out in the right size, but after uploading those tiny little postcards somehow want to be huge wall-size sheets of enlarged pixels.

That would not matter, but what really brings her down is that she can't use keywords in the upload client. She has to first select the images by keywords in the browser, export them as a group, and only then start the upload process. It would be better for her way of working to just start the upload client and select the images from there.

Cathy and computers are not a good match, she feels. But there's no going back. Steve – help!

So, universal content metadata could enhance moving content between applications. Briefly, it is essential to provide the semantics of each metadata attribute to allow application writers to use the metadata, and to allow the metadata being converted from a format to another.

One desirable aspect of universality is extensibility. The metadata format should not be carved in stone, as new content types, and new uses for old ones keep coming up. Therefore, the formats should allow adding new fields, new data types to existing fields, and perhaps redirection or even redefinition of existing fields. This may be an intractable problem; however, some existing formats (such as MARC) successfully incorporate extensions in their basic mechanisms. Our approaches to extensibility are discussed in Chapter 5.

Last and definitely not least, there is one more challenge that is not technical, but human. The challenge we are referring to is abuse. Unfortunately it seems that regardless of planning, protection, and technical solutions, you cannot underestimate the amount of work some people are willing to invest in order to create something that is harmful. There are various reasons for generating low-quality, falsified, and wrong metadata, such as ignorance, mischievousness, and greed.

Mischievousness is generally about producing metadata that may be insulting, absurd, and irrelevant intentionally. This phenomenon is actually ancient and part of human nature; small numbers of people tends to get pleasure from ruining other people's work. Also, this is something that plagues many open and shared Web services, such as Wikipedia.

Ignorance is related to the fact that people are not aware of the possibilities of the metadata and use it carelessly. In contrast, greed is an opposite case: some parties who are abusing metadata are well aware of its power. A good example of the effect of greed is that initially the Internet search engines used metadata that was embedded in the Web pages. Soon some (adult) content and service providers noticed this and started the vast abuse of the keywords in order to get more visibility to their product in search results. As a result of this abuse, the existing metadata fields of Web pages are today largely ignored by the major search engines.[43]

4.9   Yet Another Challenge: Interoperability

In order to support mobile content across multiple applications, content types, and application and device generations, the interoperability must be thoroughly considered. Metadata can help here, too. In this section, we will discuss both personal content device ecosystems and application interoperability issues, considering both applications that are used concurrently, and the need to migrate between application generations.

4.9.1   Personal Content Device Ecosystem

The consumer electronic (CE) devices offered to consumers are slowly being connected, especially those targeted at GEMS actions with personal content. There are AV amplifiers that connect to the Internet radio stations (Yamaha RX-V2700); many DVD/DivX players are capable of presenting streaming media (KiSS DP-600); and car audio system may contain a USB connector for easy listening of digital music from portable players (JVC Arsenal KD-AR770) or a cradle for Apple iPod. There are also devices that enable streaming content stored on a PC to a home stereo system or television set via a wired or wireless network (Pinnacle Showcenter 1000g). Not surprisingly, then, the development of personal content ecosystems is primarily driven by CE device manufacturers. The standards recently under development for interworking between components (such as DLNA, see below) are paving the way for truly functional device ecosystems.

We assume that the personal content device ecosystem consists of one or more mobile devices – smart phones, portable music players, and the like – and one or more mobile stationary devices – such as set top boxes and home media servers. In the device ecosystem, all components need to connect to each other to perform personal content management operations.

The simplest example of a personal content ecosystem is a PC that acts as a *central home media server*, performing back-end content management tasks. The server may be a personal computer located in

---

[43] The Net crowds have started by re-using those fields with so-called *microformats* (http://microformats.org/about/), adding descriptive keywords that are not directly visible to users but add some semantics.

Figure 4-11. "Slug" (left) and Topfield TF-5100 (right) (Photos: courtesy of Tero Hakala).

a living room, or a dedicated storage device tucked away in a cupboard, accessible through a fixed local area network (LAN) or wireless local area network (WLAN) at home. This server is used to store the content centrally, and perhaps to stream music, video, and photo presentations to a TV set. A typical case might be a family server that is accessible to all family members, and where all content common to the whole family can be conveniently stored and retrieved. Therefore, the family members are able to transfer all photos taken on a trip to Sweden, including those from mobile phones, still cameras, and video cameras. The media server may then provide a common backup service, for instance, without unnecessary duplicates.

A device worth mentioning is Linksys NSLU2 (also known as the "Slug"). Slug is a small, low cost network storage device, designed to share up to two USB HDs or flash drives to form a small network (Figure 4-11). For instance, Slug can be connected to a USB2-equipped PVR, such as Topfield TF-5100. Then Slug essentially connects the storage device to the Internet. The firmware can be easily flashed, which has resulted in vivid communities[44] around the development of a variety of different media applications for Slug.

An alternative to the "PC in the living room" scenario is keeping the PC in the cupboard, but installing a thin client-like *media streamer* beneath the TV set. The media streamer connects to the TV set and audio equipment in the living room, and provides a simple on-screen user interface for accessing the content stored on the PC. A network connection to the PC must be available, and the server must be on at all times (or at least it must be possible to wake it up through the

---

[44] See, for instance, http://www.nslu2-linux.org/wiki/Main/HomePage

network). Content adaptation is usually performed on the PC. Otherwise, this setup is similar to the media server option.

We are not only talking about traditional computers. For instance, Microsoft and its connected gaming console, Xbox 360 Live, are aiming at a central element of the home entertainment system. For instance, the users can download TV shows and movies and view them on the TV set.[45]

A device ecosystem can also be built by applying distributed techniques, where there are no central servers *per se*, but all devices are equal members in a peer-to-peer network. In such an ecosystem, some storage redundancy is required in order to keep it operational, even if some devices are disconnected.

A recent topic related to device ecosystems is *placeshifting*. Placeshiftin[46] is a technology that allows streaming video from a home TV set or PVR via a broadband Internet connection. The streamed content can then be viewed at any location where a computer display and a high-speed Internet connection are available. Images and sound from the home video equipment are transcoded and transmitted to the new viewing location over the home Internet connection.

4.9.2 Application Interoperability

In addition to the number of content files, another challenge is that there are often several applications that need to be used to access and manage the content in different ways. Especially, considering that the sheer number of different content types is rapidly increasing, it is obvious that a myriad of applications are needed to deal with them. All these utilities have different approaches and defaults, which usually requires much effort to keep the record straight, particularly when they are expected to function together as the ingredients of one's personal history and experiences.

Besides providing support for different applications, perhaps an even more influential aspect is supporting content preservation throughout its lifecycle. In other words, migrating from a legacy content management system to a new one, while still keeping all content accessible and not loosing any changes made to it, also requires standardized import and export functions.

In the long run, smooth and reliable migration is one of the most fundamental aspects of personal content management – it ensures that

---

[45] http://www.xbox.com/en-US/community/news/2006/1106-moviestv.htm
[46] Compare to *timeshifting*, which allows viewing a recorded programming at a later stage.

the memories and experiences are kept safe and sound from one generation to another. As more content forms are permanently and comprehensively transferred to the digital domain, the importance of keeping the content alive increases further.

4.9.3 Existing Solutions for Interoperability

There are several attempts in industry to standardize the use of distributed media content with various devices in a heterogeneous environment. There are many standardized organizations creating solutions in content management, working on only partially overlapping areas. In the content management area, two important consortia are Universal Plug and Play (UPnP) Forum and Digital Living Network Alliance (DLNA). UPnP forum tries to create interoperability for data transmission and also partly at the application layer. DLNA creates use-case based interoperability guidelines for application layer. Many DLNA guidelines are based on UPnP.

COMMENT: DLNA AND INTEROPERABILITY

Imagine a world in which content flows effortlessly between all of the user's devices no matter where it is created, where it is purchased from, or how it reaches the user's home. Three forces are at work today driving convergence towards this vision. First, mobile devices are becoming increasingly capable with gigabytes of storage, high resolution displays, and mega pixel cameras. Second, content is increasingly available in digital formats from many domains including broadcast, the Internet, and pre-recorded media. Finally, devices are increasingly networked with standard protocols such as 802.11, enabling PCs, TVs, and mobile devices to communicate with each other for the first time. Because the devices come from multiple industries, standards are very important to completing the picture for ubiquitous content access. An example of one important industry effort is that of the Digital Living Network Alliance (DLNA). The DLNA creates interoperability guidelines for device vendors in the mobile handheld, consumer electronics, and PC industries that address how to discover, control, transfer, and stream content between mobile and home devices. In the future, the usages will grow to include synchronizing and searching large media collections, access to the user's home from any Internet connection, and programming PVR recording schedules. With the key forces described above and the demonstrated industry momentum on interoperability, mobile devices will offer consumers an exciting way to access, control, and

```
enjoy  their  content  both  inside  and  outside  the  home,
becoming  a  key  all-in-one  device  for  mobile  media  and
control.

Jim Edwards
Principal Engineer
Digital Home Advanced Development
Intel Corporation

Gunner Danneels
Staff Engineer
Digital Home Advanced Development
Intel Corporation
```

(Jim Edwards, DLNAv2 SC chair, HNv1 SC editor (past) Gunner Danneels, DLNAv2/HNv1 SC technical contributor on mobile guidelines, etc)

There are also many other organizations trying to solve the interoperability problems, such as Internet Engineering Task Force (IETF), the body that creates and publishes Internet protocols. The World Wide Web Consortium (W3C) creates interoperability guidelines and standards for WWW applications. They do not care how the computers talk to each other or how the data is exchanged but most of W3C standards define the semantics for the data, i.e. how the data should be interpreted and what it means in a given situation.

W3C is an important standards organization for our purposes. As we mentioned earlier, W3C mainly defines semantics of the data which forms the fundamental building block for content management. If we do not know the purpose or the meaning of the content, it is impossible to create any proper methods for its management.

One standard worth mentioning, already widely adopted by several device manufacturers, allows imaging and printing devices to communicate with each other. The standard is officially referred to as "Standard of Camera & Imaging Products Association CIPA DC-001-2003 Digital Photo Solutions for Imaging Devices", commercially known as PictBridge.[47] PictBridge enables direct printing of photos from a digital still camera, without the need for a PC. All printing options, including multiple copies, index printing, setting the print size, etc., are made within the imaging device. The standard allows direct printing between any compliant still digital cameras and printer, regardless of the model or manufacturer.

---

[47] http://www.cipa.jp/english/pictbridge/

4.10 The Dream: When Metadata Really Works

In the preceding sections, we described different opportunities and challenges of metadata. As one desirable state, consider the following scenario starring Bob the activist:

THE MAGICAL CHAMONIX TOUR

Bob is kicking himself. What seemed like a simple job is turning into a nightmare. He had volunteered to compile the material from their Chamonix trip, edit it a little and publish it on their cell's site. He had not anticipated the amount of manual effort that goes into linking the relevant pieces to each other. Of course he could just dump the stuff to the site, but since they had taken all that trouble to collect the material, it would be a bit disappointing to let it just sit there, an unordered jumble of files.

For instance, it was simple enough to index the images and video by location with the help of this public domain geotagging site, but it was quite a task to enter the locations by hand, linking to those items that lacked the coordinates. Also, the creator metadata in the material was text-only. People's names were not consistent. It was Mike or Michael in the videos, O'Brannen in the images, MOBR in the blogs. Of course it was possible to relink the stuff but Bob's life started to seem too short for that.

When he vented off some steam on the task on the cell's blog, Trisha suggested that Bob talk to this guy from the University of Cambridge. They had some media analysis software that might help in the task. Bob followed the lead, found this very co-operative research group, and in a week Bob was running their Experience Engine software. That proved to be one of the most useful applications ever to land on his hard disk. The software was designed to take in a large collection of ill-structured media files, cluster the files for potential relations and invent meaningful metadata labels, and suggest how to link the files into what they called an Experience Base. Just what Bob needed.

After a night or two of crunching the material (and crashing twice), the application suggested some clusters based on time, location, and communication patterns. Bob was pleased to see that the application had divided the people into two groups: those who were there in Chamonix to experience the trip and those who stayed home and just received messages from the group. Bob had defined the various aliases of the group members, and it seemed that the software had filled in most of the creator metadata quite nicely.

On the first skim the clusters also seemed to make a lot of sense. On two occasions Bob had to shift video between episodes (why did the software associate Mont Blanc with dinner time? Ice cream cones with whipped cream on top? Not quite, Bob thought), and had to repair some weird automatic links between the images and video. Most of the time the software had beautifully found the correspondence between those images and the video clips, giving the video the proper geolocation tags as a by-product. Those blurry night time shots after some wine, raclette, and some more wine were quite hopeless, of course. Bob could not blame the software, as often he had no clue either to what those clips were about. In any case, the metadata made sense, and with the help of the group members they were able to order the material of the rowdy night into some sensible order.

Too bad that the automatically generated links between text messages, blog entries, and e-mails made almost no sense. For instance, what did "Cowbells ringing in my head" have in common with "Chocolate"? After consulting with the university researchers, Bob realized that they were lacking the proper vocabulary. They linked the material to the proper ontologies on hiking, the Alps, mountain flora, orienteering, and the like. Some of the ontologies lacked French versions, so they had to resort to another step of automatic translation between the definitions. Bob then ran the association discovery again, and the result was far better. Not perfect but already usable. To his surprise even the pictures of the signposts now turned up in right places on the map.

After a few nights polishing, Bob released the full material on the group site. He asked the people to repair any obvious blunders, and the cell jointly arrived at a useful version within a week. They then arranged a "Retour a Chamonix" night to celebrate the experience, this time with Ritchie joining the fun. They wisely had decided to skip the raclette this time to leave more room for the wine, which seemed to be an essential part of the Chamonix re-experience.

Bob, Trisha, and Gerry all presented their favourites of the hypermedia shows that Experience Engine had automatically compiled. The group agreed that the presentations were somehow strange, with unlikely structure, but still quite amusing and refreshingly different to standard slideshows. They decided to use the same recording setup in their next upcoming demonstration in Milan next December.

Functionality related to standardized metadata is a central part of applications and services as it facilitates basic functionality for personal content management with operations such as comparing, organizing, searching, and browsing.

In essence, (smart) metadata is not only plain text or thumbnail images, but links to other objects that are related in some way. The relations and associations facilitate the dawn of new applications, services, and content types that work seamlessly together. As part of the interworking, the applications and services can access any objects and they are always up-to-date and synchronized as the low-priority processes are performed, for instance, when the device is in an idle state in the background.

4.11 References

Aaltonen A., Huuskonen P., and Lehikoinen J. (2005) Context Awareness Perspectives for Mobile Personal Media. *Information Systems Management* 22 (4), September 2005, Auerbach Publications, pp. 43–55.

Baca M. (2000) *Introduction to metadata: Pathways to digital information.* Online document, available at: http://www.getty.edu/research/institute/standards/intrometadata/index.html

Cheverst K., Davies N., Mitchell K., and Efstratiou C. (2001) Using Context as a Crystal Ball: Rewards and Pitfalls. *Personal and Ubiquitous Computing* 5(1), February 2001, Springer-Verlag, pp. 8–11.

Dey A.K. (2001) Understanding and Using Context, *Personal and Ubiquitous Computing* **5**(1), Springer Verlag, 2001, pp. 4–7.

Dovey M.J. (2000). "Stuff" about "Stuff" – the different meanings of metadata. *VINE* 116, January 2000, pp. 6–13.

El-Sherbini M. (2000) Metadata and the future of cataloging. *Library Computing*, Volume 19, Number 3/4, 2000, pp. 180–191.

ePerSpace (2004) *ePerSpace – IST integrated project. Report of State of the Art in Personalisation. Common Framework.* 2004. Online document, available at http://www.ist-eperspace.org/deliverables/D5.1.pdf

Gilliland-Swetland A.J. (2000) Setting the state: Defining metadata. In: Baca M. (ed) *Introduction to metadata: Pathways to digital information.* On-line document, available at: http://www.getty.edu/research/institute/standards/intrometadata/index.html

Jenkins H. (1992).*Textual Poachers. Television Fans and Participatory Culture.* Routledge, New York and London.

Kennedy J. and Schauder C. (1998) *Records Management a Guide to Corporate Record Keeping.* Addison-Wesley-Longman, Reading, MA.

Lehikoinen J., Salminen I., Aaltonen A., Huuskonen P., and Kaario J. (2006) Meta-searches in peer-to-peer networks. *Personal and Ubiquitous Computing*, 2006, Number 10, pp. 357–367. Springer Verlag.

Manjunath B.S., Salembier P., and Sikora T. (eds) (2002) *Introduction to MPEG-7: Multimedia Content Description Interface.* Wiley & Sons, April 2002.

Mathes A. (2004) Folksonomies-Cooperative Classification and Communication through Shared Metadata. Online document, available at http://blog.namics.com/archives/2005/Folksonomies_Cooperative_Classification.pdf

Rosenblatt B. (2002) XMP: The Path to Metadata Salvation? *The Seybold Report*, Volume 2, Number 16 (November 25, 2002), pp. 3–7.

Smith J.R. and Schirling P. (2006) Metadata standards roundup. *IEEE Multimedia*, Volume13, Number 2, pp. 84–88, April–June 2006.

Tešić J. (2005) Metadata practices for consumer photos. *IEEE MultiMedia*, July–September 2005, pp. 86–92.

Chapter 5: Realizing a Metadata Framework

application attribute context
device file framework
image metadata object
ontology relationship schema value

In the human mind, everything is connected. We tend to memorize and process information through associations, no matter how vague or incomplete they may be. Computers, on the contrary, are precise, so even if a single letter or number is not exactly right, they refuse to understand. This makes co-operation between a human and a computer so difficult that they literally do not understand each other.

In this chapter, we do not try to teach a computer to understand humans, but rather to change some of its internal structures to match some aspects of human behaviour, eventually making it support a more human way of thinking. The goal is to create mechanisms that support, among other tasks, organizing and finding content by using associations.

We have already discussed many issues related to personal content experience and management, and pointed out that metadata is a key

Personal Content Experience: Managing Digital Life in the Mobile Age J. Lehikoinen, A. Aaltonen, P. Huuskonen and I. Salminen © 2007 John Wiley & Sons, Ltd

enabler in solving them. Upon inspecting this solution more thoroughly, we soon notice that the quest for reaching enjoyable personal content experience is not over. Metadata, the much welcome solution, presents a new breed of challenges. Now the question is, how to solve this metadata challenge – how to create, store, access and, finally *manage* metadata.[1]

In this chapter, we discuss metadata management in more detail, starting with an overview of *content* management in mobile devices and drawing from there the requirements for *metadata* management. Finally, we describe a prototype of a generic content management framework available for all applications, first introduced in Salminen et al. (2005). With our prototype, we specifically target mobile devices, yet emphasize that the solutions presented are by no means exclusive to mobile devices but are easily ported to stationary devices in applicable parts.

5.1   Metadata is a Solution . . . and a Problem

There are four terms above others in this chapter: *content, context, metadata* and *ontology*. Content is always the ultimate reason for the existence of the framework. It provides access to the bits and pieces that the user Gets, Enjoys, Maintains, and Shares.

Context, as discussed in section 4.7, provides additional information related to the situation, be it at the moment of taking a photo, or viewing it later. It puts the content into the bigger picture by incorporating attributes such as location, weather, nearby friends, goals of current task, and so on. Context is usually of a dynamic nature and so constantly changing.

Metadata, then, describes the content object, as discussed more thoroughly in Chapter 4. There is a close relationship between context and metadata, since context information is in many cases "frozen" and stored as metadata. For example, when you shoot a photo, the current location (which is a part of the context) is stored as metadata and associated with the photo.

Finally, ontology is a term used extensively towards the end of this chapter when we describe our framework in more detail. For now, it

---

[1] Logically, if metadata solves the data problem, then meta-metadata should solve the metadata problem. As we shall see, meta-metadata plays an important role in our metadata management framework. However, to solve meta-metadata problem, we do not intend to apply meta-meta-metadata.

is sufficient to know that ontology is basically a list of items and their relationships that our framework knows to exist. It will be discussed in more detail in section 5.6.

For the purposes of providing the applications with access to content, context, and metadata, we have designed and implemented a personal content framework. The developed framework acts as a real-world case study, showing how to deal practically with the issues related to content management, and how the application development is affected. The framework provides all application developers with a good, strong, and level scaffolding on which to build robust applications using metadata, and offers a unified way of storing and managing metadata regardless of its type or origin. Furthermore, it allows developers to focus on how to leverage metadata for creating better applications, instead of having to deal with low-level metadata management issues.

As discussed throughout this book, the importance of metadata increases along with the amount and variety of content objects. As pointed out in Chapter 2, even mobile devices with a limited storage capacity are equipped with gigabytes of cheap storage space, and the amount of files is close to that found on PCs a decade ago. For example, as recently as 2006, a 1 GB memory card was the norm for storage in mobile devices, and 2 GB cards are now available and not much more expensive. This means that the amount of available storage memory in mobile devices will continue to increase rapidly. At the same time, mobile devices have expanded from mere "E" devices into "GE". You can shoot photos and videos, record audio, create multimedia presentations, and so on. A mobile device can also connect to your home network and access all content you have on your media servers and in other devices in the personal content device ecosystem. Finally, the connectivity allows easy and instant sharing.

All the facts described above add to the number of accessible content objects from within a mobile device. As a consequence, traditional directory or folder hierarchies are not adequate for maintaining or even finding your own content (section 3.3). Then metadata becomes a key element in all aspects of content management and so also needs to be managed.

From a personal content software point of view, there are two primary options for metadata management. Either each application takes care of its own metadata by using its own designs and methods, or the operating system (OS) provides a system-wide metadata management service that takes care of metadata for all applications. Both approaches have their pros and cons. Since we emphasize the importance of metadata in all content management tasks, we choose the latter option and state that *metadata management is a vital part of the operating system* as long as content-intensive applications are concerned.

Some issues addressed in this chapter are common to those dealt with in traditional Content Management Systems (CMSs). Before more fully discussing the metadata framework, let us look at some of the problems with content management, especially those caused by mobility. For more in-depth discussion on the technological issues and practices related to CMSs, see Boiko (2005).

5.2   Challenges in Distributed Mobile Content Management

Mobility implies that the content is always with the user, regardless of the time and place. This results in what is perhaps the most fundamental issue with mobile personal content, especially considering the remarkable growth rate. Not all personal content can be stored in the mobile device, due to limitations in storage capacity, wireless network capabilities, and media processing capabilities. Mobile content management is by default distributed content management. This aspect determines many characteristics of mobile personal content.[2]

5.2.1 Storage

An important aspect associated with mobile personal content is the device ecosystem, discussed in Chapter 4. In essence, it is a question of owning several devices, mobile or otherwise, that are capable of storing and providing access to personal content. As a consequence, personal content is probably spread across the device ecosystem.

The most obvious question related to distributed content is to determine where the content has been stored, if not in the mobile device. There are three obvious options available:

1. All content is stored in the mobile device.

2. A subset of content is included in the mobile device, and the rest is elsewhere (PC, set-top-box, Internet, etc).

---

[2] Yes, mobile storage is increasing, but so are the sizes of content files and their number. We believe that mobile devices will, in the future, only be able to hold a fraction of the personal content – except maybe for people with limited content demands.

3. A subset of content is included, and the rest of the content is indexed so that its location can be determined, also when mobile.

Let us now consider each option in more detail.

5.2.1.1 All Content in a Mobile Device

Currently, it is not realistic to assume that all content will be stored and used in a mobile device only. Most likely other content storage formats and types, such as hard disks in PCs, DVDs, memory cards, and PVRs also contain loads of personal content. Therefore, even if all content is stored in a mobile device, a copy of at least a subset of it may also be stored elsewhere or at least used with other devices. However, we can confidently assume that a significant portion, if not most of personally *created* content, will be stored (at least temporarily) in mobile devices, since it was created by them in the first place. This, in turn, implies huge challenges in keeping the separate collections and devices in sync. Synchronization will be discussed in more detail below.

5.2.1.2 A Subset of Content in Mobile Device

In this case, the mobile device contains some subset of the user's personal content, but is not aware of any other content in any other device, so no synchronization is required. A typical example might be photos taken with a mobile phone's built-in camera. They sit firmly in the phone's memory card. A special case is a digital camera, which only temporarily stores the images, until transferred to a permanent storage device, after which the camera's memory card is wiped clean. The same is also true for most mobile devices targeted towards fitness, such as wristop computers with heart rate measuring capability (section 7.5).

5.2.1.3 A Subset of Content in Mobile Device, The Rest is Indexed

This case differs from the above in the sense that even though the mobile device does not have all content, it is aware of its existence and has a link to it and capability to retrieve it when needed. As a result, synchronization is needed in order to provide the mobile device with information on this content available elsewhere.

A fourth option can be extrapolated from the above: some content is only in mobile devices (for instance, SMS), some are occasionally transferred to PC (images), some are updated (MP3), some are synchronized (contacts), and some are searched from the device ecosystem.

Another aspect related to storage is the system that is used behind the scenes. Is a file system sufficient, or would a media database be needed?[3] There are some trade-offs between the choices.

5.2.1.4 A File System or a DBMS?

Speaking of legacy and interoperability, a file system is a safe option, as next to all computing systems can manage files. Opening, editing, and transferring files are standard operations that we take for granted. Furthermore, a file system does not, at least in theory, require any administration, unlike several database management systems (DBMSs). Obviously, a zero-administration database system is the only viable choice for consumer-based products.

One aspect to consider is efficiency related to searching. Files can be indexed and searched for, as described above. However, databases, relational or object-oriented, are far more efficient and sophisticated in supporting complex searches, different views to information, and so forth. The same holds for most operations that need to be performed on content, such as backups and synchronization. Such functionality is built into most DBMSs.

Another issue is privacy and metadata management. With file-based systems, all additional information related to a file needs to be included in the file itself. The metadata can be stored in a separate file, and the content file will just contain a reference to it. Such a method is extremely error-prone, since it is all too easy to separate the files from each other (see section 5.5 on metadata manifests). This may also include sensitive information that is private in nature. Should a file with such information end up in wrong hands, severe damage may follow. A good example is the recent inclusion of unwanted metadata in some public documents.[4]

With a DBMS, metadata can be stored in the database, separately from the content itself. When a piece of content needs to be exported as a file, a policy can be applied to determine what (if any) metadata should be included in the file that is to be exported. Better still, in case the recipient has a similar DBMS in use, the content can be replicated to the recipient's DBMS with defined security and privacy options, such as what metadata is transferred in the process.

---

[3] A database is also implemented as a file (or a collection of files), since files are where the computers store data in the first place. However, accessing a database (from a user's point of view) and the way it is organized internally is inherently different from a file system.

[4] http://www.theregister.co.uk/2004/02/02/microsoft_releases_metadata_removal_tool/

5.2.2 Synchronization

Synchronization will be one of the most prominent challenges in developing robust content management systems, even more so when mobile devices are involved. Synchronization itself is a vast domain, discussed in detail in, e.g., Bernstein et al. (1987); Adjeroh and Nwosu (1997); and Lin (2002). In this section, we do not present any detailed technical solutions, which would be beyond the scope of this book, but confine ourselves to discussing the synchronization-related challenges that the developer of a personal content management system will eventually have to face and solve, often on a case-by-case basis.

Currently, personal content is stored and used in a myriad of different devices that may or may not connect to each other. Consider the implications to the user: how to create a shared family photo album, when each family member has a digital camera of their own and transfers the images to their own laptop computer? How to synchronize favourite playlists between a home media server and a portable music player? Sometimes an automatic synchronization based on, for instance, the play count, can be applied, whereas sometimes the user may prefer to synchronize manually created playlists.

Synchronization implies keeping track of changes made in the data store, such as additions, deletions, and modifications. Occasional conflicts need to be addressed, for instance when the same object has been modified at both ends. In those cases, the system needs to resolve the conflict, either by asking the user explicitly for instructions or by applying predefined rules.

A typical example of explicit synchronization, from a consumer point of view, is switching from an old phone to a new one. The need to transfer all locally stored personal content to the new device is obvious and essential. Typically, there are two options available. Either the contents are first transferred to a PC, which is then used to copy the contents further to the other device. Another solution is a direct point-to-point connection, where the content is transferred directly between devices.

What is perhaps the most challenging aspect of synchronization related to personal content is keeping the user from having to deal with it. Deciding once or twice which version of a photo to carry in a mobile device could be considered fine, especially if the user interface for such an operation is properly designed. However, on a general basis the user should be able to happily forget the problem of synchronization altogether. This is a goal not trivial to achieve.

One of the desirable features of a connected mobile phone is a synchronized clock, since the time of day is provided by the operator.

This removes one of the key challenges in synchronization. In fact, a large number of unconnected devices will have their clocks set wrong.[5] Lately, with the advent of Internet connected devices, this problem has lessened but new problems have arisen. Many new devices are capable of synchronizing their time to some known reference time, by mostly using Network Time Protocol[6] (NTP) or Simple NTP (SNTP). One manufacturer of a popular router used the NTP server at the University of Wisconsin to set its time. Since all routers from this manu-facturer, from all over the world, connected to the same Wisconsin server, the server became quickly overwhelmed and could no longer serve anyone. For more details about the incident, see Mills et al. (2004).[7] Consequently, timestamp-based synchronization is prone to error. It is difficult to ensure synchronized clocks in a distributed environment.

5.2.3 Version Control

"Version control" is a term often associated with software develop-ment. It allows keeping track of changes made to the source files, probably by multiple software engineers, and making different builds of the whole project by choosing which versions are to be included. In addition to software projects, version control is also an essential component of personal content management.

Often, there are many versions, or instances, of a piece of personal content. For example, the user may have several versions of the same photo: the original, one with adjusted colour levels, and one scaled down for web use. Likewise, a song digitized from an original analogue recording may be stored on a CD as an uncompressed version, while a compressed version may exist for music collections. Furthermore, a new version is implicitly created when an image is sent to a friend. What is important to consider with the different versions is to deter-mine whether the subsequent changes made to the derived content are also reflected back to the original. For example, when the photo album software in a PC keeps track of how many times a certain photo has been viewed full-screen, should the viewing count for a version of the same image in a mobile device be included as well?

---

[5] This is the mobile version of the infamous blinking "00:00" clock problem commonly seen in VCRs (Video Cassette Recorders) and microwave ovens. Setting the VCR is too complicated for many people, so they just leave the clock unset.

[6] http://www.ntp.org

[7] http://www.cs.wisc.edu/~plonka/netgear-sntp/, checked 15.12.2006

A deeper challenge is due to the *changes made in the content objects themselves*. Resolving the issues that arise requires both synchronization and version control.

This leads us to philosophical issues, such as defining the similarity of two objects. For instance, when a photo is edited for colour balance, is it the same as the original version? In case the original file is replaced in the process, it obviously is the same from the storage system's point of view. However, as far as synchronization is concerned, what should be done with the original photo stored in a mobile device, and the edited version in the desktop computer? Should the original be deleted? From the user's point of view, this may or may not be the case.

A partial solution is to delete nothing until explicitly instructed so by the user, but to keep control of all changes and allowing an indefinite number of undo operations. The problem in these cases is the ever more rapidly growing amount of content, and so finding the suitable version of a photo for a slideshow, for instance.

5.2.4 Backing Up

Important and valuable personal belongings have always been kept safe by the means of safes, safe deposits at banks, etc. Digital content is, or should be, no exception. When discussing the features of individual content objects, one of the most important of them is, not surprisingly, personality. In many cases, content has not become personal for free: in order to create or personalize it, you may have had to put in some effort in doing so. If the content is lost or damaged, the changes made by you are also lost; there is no authority that can restore it. The solution is storing the content, more or less frequently, to an alternative storage medium. Backing up data is a typical application of synchronization.

ONE SERVER TO RULE THEM ALL

Angie thinks her dad has gone nuts. He had bought yet another network disk for their home network. As if they would not have enough storage already, with four computers in the house – Dad's Mac, the media centre Mac in the living room, Adam's game PC, and Angie's own laptop. Everyone had access to each other's computer's files, although Angie carefully kept her own personal files away from her public folder.

They had devised a cross-backup strategy: all their videos, images, and music were stored in the media centre and backed up in the office Mac. In addition, her dad sometimes burned DVDs of the family albums and sent copies to Uncle

Ron. Angie maintained her drama club web pages just on her own machine, but they were backed up in the media centre, and so on. Nobody would touch her brother's gaming machine for fear of getting infected. Dad insisted that Adam use virus scanners, which he did, but Angie suspected that he sometimes switched that off to avoid trouble with game installations.

Now her dad had the idea of collecting all their family content on one server that would then allow universal access from the Macs and the PCs via file systems and local WWW browsing. That sounded overly complicated, but Dad explained that would give access to any file from any machine at any time. The individual machines would not have been on all the time, and so on. Oh, and Granny could hook into the server too through her browser. Angie was not so sure about that . . . She just let Dad rant on about his plans. She didn't care so much about it after all. As long as she could keep her own MP3s to herself and not let Adam hack them.

A backup in its most simple form is just a copy of the piece of content, preferably stored separately from the original. Alternatives include online services that take care of backups for a monthly fee, such as Carbonite,[8] or distributed peer-to-peer systems where the content is stored redundantly in several computers around a serverless network (Dingledine et al. 2001).

5.2.5 Content Adaptation

Personal content can be enjoyed on a myriad of different devices. Some are equipped with large screens, whereas some do not have a screen. Some are able to reproduce music in its finest detail, whereas some are practically mute. In order to cope with devices with such a huge capability range, one must consider how each content object can best be enjoyed with a certain device. Reacting to such changes by altering the content itself is referred to as content adaptation (Mohan et al. 1999; Katz 2002).

The first aspect to consider is *device capability*. Does it have a pointing device? Can it output sound? It is the responsibility of the device to inform its capabilities to the content provider. The content provider, then, can determine what content and in which form can be transferred to the requesting device. When the device capabilities allow, say, displaying images, but only contains 64 000 pixels, it is usually not a good idea to transfer a four-megapixel image to this device for mere viewing purposes. Rather, the content should be adapted to the capabilities of the receiving device. This process may include reducing resolution or bitrate, or increasing the compression ratio, for instance.

---

[8] http://www.carbonite.com/default.aspx?

THREE DEE

Cathy was a little envious. Deena was so cool with tech. Not that Cathy would really lust for the gadgets, mind you, but it was disturbing to see how easily Dee could get into the latest functions of whatever gadget she found – as long as it was useful for her art. Deena was so driven, so deterministic that way. Whatever was needed to realize her artistic visions, she would do it, even if she needed a PhD in quantum physics for the purpose.

Now, Deena had devised a 3D model of a renaissance building. It was complete with the furniture, paintings, wallpapers, ornaments, everything. Except for people; Deena was not into people. Buildings and atmospheres were more her thing, and her crazy surrealistic scifi effects, which in an eerily appropriate way contrasted the historic settings. Flowing wallpapers were really creepy! Anyway, they looked at the model on Deena's computer, and she explained how her editor software produced various versions of the model, even one in normal compressed video for people who lacked 3D hardware in their phones.

They tried viewing the building on Cathy's old trusty phone. It was not fun, as the download took ages, and the resulting playback was jerky, but still it could be done. On Deena's phone – not the latest model either – viewing was already quite enjoyable. The initial download wait was long but the model was smoothly animated once it got there. Anyway, Deena explained how she planned to render a string quartet in one room such that if you lacked channel capacity, you could just hear the music and see a still frame portraying the orchestra. If you had a better link you would get the whole 3D works in surround sound.

Cathy was not sure if she liked such adaptations or not. In her art, she wanted to portray her subjects (these days, herbs and flowers) as thoroughly as possible. If the viewer could not get the hires pics, maybe it would be better to abstain completely. Or maybe not. There was some business building up on their web site, and it might be useful to include some self-adaptive content. But would their server allow that? Steve – help!

However, the characteristics that affect content adaptation are not restricted to the receiving device's capabilities. Another factor is *network capability*. Estimating the time required to transmit a piece of content is determined by not only the network capability, but also current load. As pointed out in Chapter 2, network speed will remain the bottleneck in content transfer in the foreseeable future. However, it can be argued that the increasing compression capability of recent media codecs, such as Advanced Video Coding (AVC) (Marpe et al. 2006), move some of the burden to the processor while easing the requirements of the network throughput.

What still needs to be considered is the *intended use* and *desired quality level*. For instance, transferring a full 5 Mpix digital photo is not sensible if the photo will be used in a web page only, whereas for printing, a full copy is preferred even though the receiving device may not be able to show the full detail.

Similarly, the *devices accessible nearby* could affect the need for adaptation. For instance, instead of a device's tiny screen, one might show a video clip on a larger screen by connecting the mobile device to a TV set. In this case, even though the screen resolution of the device itself might justify content adaptation, the intended usage suggests otherwise.

To summarize, content adaptation is far from a trivial issue, yet in many cases it affects the perceived usability and user experience a great deal.

5.2.6 Locating the Desired Piece of Content

To enjoy a piece of content, the user needs to first locate it. Theoretically, there are two methods for finding a desired piece of information: either by browsing, or by searching (although browsing can also be understood as a form of visual searching). Currently, searching is often associated with search engines, such as Google,[9] that are used for finding material on the Web. However, as the amount of content stored in local devices rapidly increases, search tools for personal content will be increasingly required. One such tool is Google Desktop that indexes all local content upon installation on a PC and, once installed, all subsequently added content is also indexed. Then, it can be used as a keyword-based search tool for locating not only remotely, but also locally stored content.

As far as file systems are concerned, searching in its simplest form is based on *file names*: you provide some keywords, and the system scans through the file system, comparing the keywords to the file names. Each hit is considered a result.

This approach has severe limitations, especially related to file management versus content management (section 3.1). First, file names do not usually convey any information related to the content of the file, or do so only partially. Worse, many devices and applications that are capable of creating personal content apply an automatic naming scheme to all created files (such as a prefix and an index number). "image001.jpg" and "imag002.jpg" are as usable to the file system as are "mum and kids.jpg" or "my biggest fish ever.jpg". After all, the file name is used primarily for internal identification by the file system.

File names are just one form of document descriptors. This leads us to *keyword searching*, where a set of keywords[10] are associated with

---

[9] http://www.google.com/

[10] That is, metadata. For more discussion on metadata, refer to Chapter 4.

a document and used as document descriptors. Again, the search query consists of keywords. These systems work best when the keywords used are highly selective, that is, they occur infrequently in the documents in general, but are common in the documents that the searcher wishes to retrieve. In order to be effective, keyword-based searches require enough relevant keywords in each document descriptor.

Another simple search is based on creation and modification time. All file systems keep track of time when a file is created or modified. So it is possible, for example, to search for files that were created today or yesterday. However, the usefulness of timestamps is directly comparable to how well clocks are kept synchronized (as discussed above).

An extension to keyword searching is *concept searching*. A thesaurus is used to augment the original query with more potential keywords. Also grammatical and morphological knowledge may be applied in the search, such as plurals for instance. Additional extensions include Boolean searching (using logical operators AND, OR, and NOT); and weighted searching (providing weights to each keyword).

Full-text searching differs from the keyword-based searches in that instead of indexing only a set of pre-defined keywords, the whole text in the document is indexed. This implies that the searcher does not have to worry about whether a certain keyword is indexed or not. Also, the full-text indexing indexes numbers, dates, etc.

Relevance feedback uses documents themselves, instead of keywords, as query terms. Hence, relevance feedback may be referred to as query by example. Once a relevant document is found, the user may ask the search engine to search for more documents like it. In this search, the search engine may use any search technique described above. This is the greatest disadvantage of this search technique – the searcher may not know which characteristics of the current document are used in the subsequent searches. This technique is best when the user is unable to find more relevant documents (Raymond 1998).

The discussion above concerns only textual information. How, then, should one search for content items that are stored in binary formats, such as music and photos? The content itself should in one way or another be used in searching.

Searching for features in multimedia files is referred to as *content-based retrieval* (CBR) (Hirata and Kato 1992; Tseng 1999). In CBR, the idea is to analyse the content, for instance identifying shapes in a photo or rhythm in a song. These features can then be searched for, either by providing examples of the patterns or features that are desired (this technique is one form of "query by example" technique), or by translating the extracted features into keywords. In this case, all techniques available for textual searching can be subsequently applied.

In many cases, CBR poses further challenges to user interface design (section 6.5). What is especially interesting in CBR is the way in which the search criteria are provided. This may be relatively straightforward with visual information, but how is one to provide an example of a specific kind of heart rate curve? Or a saved game, for that matter?

Yet another aspect related to searching is the searching range. This is especially relevant to mobile use with distributed content; specifying whether the content is searched for in the device only, from the whole personal device ecosystem, or from the Internet, heavily affects the performance and thus perceived efficiency of the whole system. Furthermore, performing a search on a peer-to-peer network, where the results indicate currently available content, is inherently different compared to searches on an indexed database (such as current web searches), which may contain outdated information on content availability.

5.3   Different Types of Metadata

Prior to diving into the mysteries of metadata management, some discussion on metadata classification is in order. Next, we will introduce our way of classifying metadata.

Metadata is not just something that is glued onto the data it is describing, but is inherent, built in every content object, system component, and piece of software in a computer device. Metadata related to a content object is as much a living thing as the object itself.

Usually we regard metadata as simple attributes or tags that describe the content object to some extent. For example, all files have a creation date and a name attribute, providing information on when the file was created and how it can be referred to in the system, respectively. Other examples of tag-like metadata are album and artist information in MP3 files, or the aspect ratio of a feature film on a DVD. What else would one need to gain a sufficient access to and control over their personal content?

We have expanded the concept of metadata far beyond mere tags, as our four categories of metadata illustrate:

1. Attributes, that is, tags

2. Context information

3. Associations or relationships between content objects

4. Usage history or events.

Recall the discussion related to derived metadata (section 4.4), where derived metadata seeks to increase the abstraction level of the content. Of the above, attributes are metadata, whereas all the other classes are derived metadata.

We see context information as an important source of metadata, especially as far as mobile use is concerned. Context usually does not describe the content object itself, but the situation where it was created or enjoyed. Context information plays an important role when creating automatic collections and playlists and trying to find lost content objects. Section 4.7 provides more discussion on how the contextual metadata benefits the user.

Associations are the kind of metadata that requires more than one content object. They describe how two or more objects relate to each other. A whole section about relationships will follow, as we consider them to be the most important form of metadata.

Usage history is metadata describing how the content has been used and interacted with. It includes use count history, such as the playcount for a song; it also reveals when an image was edited, or how many multimedia messages were sent to a certain recipient. Among other tasks, usage metadata can determine what is the most important (read: most often used) content object in a mobile device.

It is not always trivial to assign a given piece of metadata to one of the categories listed above, yet it is not particularly important, either. This division is included to show that a wide range of metadata, with a rich set of characteristics, exists beyond mere tags.

In addition to media objects, applications also have metadata, such as version information, author, and compilation date. Metadata is generally understood as data about data; in this particular case, it is data about an application.

5.3.1 Tags

Most common metadata are simple attributes, that is, tags. Some tags have more or less loosely defined semantics, such as ID3 tags for music files or IPTC tags for images. These attributes are often called (key, value) attributes since they are used in pairs. An example

ID3 (key, value) pair can be expressed as *"artist = Schbert"*. The key, here "artist", defines meaning for the value, here "Schbert", saying that Schbert is the creator of the song.

Recently, another way of using tags without clearly defined semantics has arisen. These so-called *folksonomies*, are tags without a key or rules on how to use them (section 4.1). Schbert's song "But not today," could be expressed simply by listing of plain tags, i.e., just values without keys: Schbert, Pest of Schbert, 2006. There is no way to understand what each tag means but usually music listeners can quickly decipher that "But not today" is a song by Schbert and it is in the album Pest of Schbert released in 2006. Folksonomies are extensively used in services like YouTube[11] or Flickr,[12] where consumers create and tag the content themselves. Users see what tags others use and start using the same tags themselves to describe the same properties. Usually, these services also automatically link content with the same tags to each other, thus creating groups of similar content.

Folksonomies are easy and flexible to use since the user can choose whatever tags he or she likes or thinks describes the content best. They are useful for humans but extremely difficult for computers, since there is no predefined semantics behind them. We believe that for efficient metadata management both approaches, i.e., clearly defined semantics and freedom to create your own tags, are crucial and all metadata management frameworks must support them.

5.3.2 Context Capture

You do not usually think much about metadata while taking your vacation photos, even though metadata plays such an important role. At the click of a button you freeze the image in front of the camera, but the visual image contains only a tiny part of the prevailing reality. There may be friends hanging around, yet not all may be visible in the photo. Furthermore, the image does not usually identify the location, or the name of the mountain you use as a backdrop. All that information, generally called context information, is usually lost for good. While it is important to capture the images or sounds, for the purpose of re-living the experience it is equally vital to capture the whole situation, too.

As pointed out in section 4.7, there are a number of potential context information sources, some of which are easier to obtain than others (Figure 5-1). Our framework stores available contextual

---

[11] http://www.youtube.com/
[12] http://www.flickr.com/

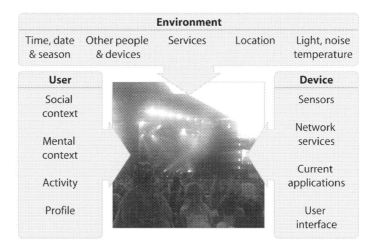

Figure 5-1. Different sources for context information.

Figure 5-2. Context as metadata.

information, including location as GPS coordinates or the network cell ID, all detected Bluetooth IDs, the current song playing in the media player, a link to the calendar entry for the day, and battery and network field strengths. The purpose is to capture as much as possible of the situation and associate it with the content object. We call storing the situation *capturing the context as metadata* for the new content object. To us, any framework managing metadata must also be able to capture and store context.

The importance of capturing the context becomes evident when enjoying the content. It is difficult to organize images, music, and other media by using their content as a key. For example, grouping all photos by the people (that is, analyzing the visual content and detecting and recognizing faces) would be a nice feature, but is beyond currently available technology. You would probably end up having photos of your wife appearing together with images of cows on pasture, and although that might lead to some interesting conversations, it is probably not what you want. With the help of context information, a rich array of grouping and organization schemes can be applied. The recorded Bluetooth IDs, for instance, can provide some hint towards people nearby. In essence, context information can be used to create automatic relationships between the content created or received in the same context.

THE 60 PENCE LOOT

Angie wants to show Millie her good close-up pic of 60 Pence. Millie is for some mysterious reason quite into the music of that fatso rappa, and she is so pissed that she was not able to come to the Hyde Park show due to a funeral. Angie does not care for that gangsta too much, but she might as well share the pic she took by the stage – only if she could find it.

   There are some 40 000 pics on Angie's mobile. Looks like that good pic she took is simply not there. Where has it gone? Anyway, there are lots of others. Angie selects the "60 Pence Loot" collection that Eddie sent her. He had collected all content that he and his friends had bootlegged at that concert, complete with location and other context information ripped from the concert arranger's site. Poor pics though, hers was better. The videos are ok, however, and the MP3 links in them come in handy.

   Angie copies the context parameters from the Loot collection and uses that to query for related pictures. Ok . . . these are hers, with matching contextzz . . . right, there's the one. That 60 pence guy is one big fat badass. But who carez. Yo Millie, you ready for upload?

With context recall, Angie can find all music from Eddie, and then do some more filtering to find all files that have links to 60 pence's concert entry in the diary. You do not need to type any keywords or try to remember what, if any, keywords are in the music files, but use a very natural way of browsing through associations between events, people, and files.

5.3.3 Relationships

Today, most applications use metadata as simple attributes or as tags describing the content, even though metadata can be used in much richer ways. We also want to use metadata to show how two or more items are related to each other (Figure 5-3).

In the use case above, all songs, videos, and images from Eddie have a relationship with each other. Storing that relationship as a metadata implies that you can later access them as a group. Or, if you come across one of them, you can quickly locate the rest. This allows *associative browsing*, accessing content objects with the help of other objects making searching much easier, as discussed at the beginning of this chapter. You do not have to remember keywords, tags, or any arbitrary text, but only locate one significant content object, be it an event in the calendar, a received e-mail, or an image. This is a human way of accessing objects, therefore one of the most important requirements for the framework is to be able to store and search for any kind of arbitrary connections between content objects.

Since there are many different kinds of relationships, our framework must support the addition of new relationships when needed. Thus, in addition to the standard relationships we have defined in our ontology, anyone can create their own types of relationships. The meaning

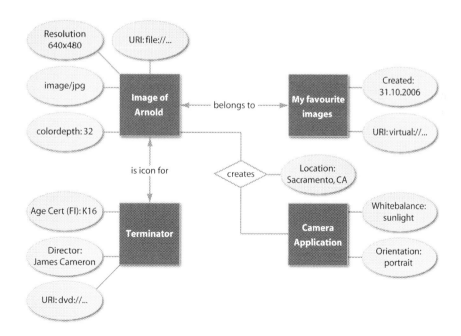

Figure 5-3. Relationships link content objects to each other.

of such a relationship, though, is evident only to its creator unless he or she publishes documentation about its semantics and thus leading to its wider adoption.

The advantage of this approach is that Eddie's loot may have its own metadata like updated date, passwords for access control, and so forth. Even if there is no physical file or object in the system as such, it still can have its own metadata. The main use of such relationships is for creating virtual objects. These objects exist purely in our framework, have metadata, and can play a part in relationships but they contain no files or bits. Thus, virtual objects are pure metadata. As can be seen, anything can have attached metadata, and in the case of virtual objects, even nothing can have metadata.

Simply put, relationships are two-way links between content objects. They show that two or more objects have a common character. It does not matter what kind of objects they are, or whether there is a single application that can handle both types. As an example, you may want to add a link from a calendar event to a note describing the meeting, a link to the phonebook for participants' phone numbers, and a link to an on-line game to be played at the event. The framework allows linking anything to everything and whether it makes sense or not is completely up to the creator of the link. The framework happily accepts and stores it for further reference.

To achieve application interoperability and common metadata usage, we do have defined semantics for some of the most common relationships. If a developer decides to use those well-defined relationships, it is assumed that they follow these definitions. However, the framework has little means by which to enforce doing so. It is totally possible but probably disastrous to use, for example, the *replaces* relationship to link two or more holiday trip photos to each other. However, some application that does automatic housekeeping or version controlling might use that relationship as an indication that it is allowed to remove other images if storage space is limited. As we discussed earlier, automatic maintaining of backups and synchronization is especially important in mobile devices where the storage space is always limited. We have defined semantics of relationships just for ensuring interoperability between applications so they do know what the intentions of other applications are.

One of the most useful relationships is the *"is used by"* relationship, which conveys that a content object is used by another one for some purpose. The relationship itself does not convey *how* the other object is used. That is left to the application developer but since all relationships are automatically two-way relationships, the used object also has metadata that it is being used. A practical example could be using several video clips, audio tracks, and still images for creating a video presentation. The *"is used by"* relationship is used as a reference count, meaning that it shows how many times an object is used by other objects. Then, while editing the video, you may decide that some clip is not needed, and consequently remove it. In such a case, the clip is not ultimately deleted, but only the relationship with the presentation is removed.

Similarly, if you try and delete an included clip using some other application, such as a file browser, the system warns that deleting the clip will have adverse consequences for the video presentation. *Smart deletion* marks the video clip to be deleted later, but does not delete the actual content until its reference count becomes zero (that is, it is no longer used by any other content object, including any application).

In Eddie's loot example, all items Eddie has sent are related to each other by a common character – they are received from Eddie. Since Eddie sends lots of stuff, gigabytes of MP3s, and video clips, it would mean that all items would have a relationship with all other items in the loot. If there were only three or four MP3 files, it would be fine. It would mean that there are six (three items having two relationships each) or twelve (four items having three relationships) associations between all of them, which would still be easily managed. However, if the number of items increases, the number of relationships increases exponentially, and finally the whole system will be drowned by all the relationships. Therefore, we need to have a method to limit the increase of the relationships.

We conclude that all relationships are two-way in nature, implying that if the video clip "jumping cat.avi" is used by the movie "all cats. avi", it automatically means that movie "all cats.avi" uses "jumping cat. avi". This already cuts the number of relationships in half, since we need to store the relationship only once per pair. Another, better way, to limit relationships is to use those virtual objects that are pure meta-data. Our framework does not enforce it but it is much more efficient to create a virtual object, here "Eddie's loot" that has one relationship between all items in the loot. Now the number of relationships increases linearly and what is even better is that Eddie's loot can have its own metadata.

5.3.4 Usage History and Events

An *event* occurs when a content object is used in any way. It can be a modification event, such as the creation of a file, or a non-modification event, such as playing back a video clip. Some events can be triggered by the system, such as receiving a message or deleting a file, while many events depend on the applications being properly triggered by the framework, for instance viewing an object. Events are used to create interaction history and usage logs.

The simplest forms of using logs include creating dynamically updated and adapted playlists, such as the most often played songs. The logs can also be used by more advanced harvesters to infer totally new type of metadata, such as when the user is alone she likes to listen to classical music, but with her friends she prefers rock. Events are also useful when making more complex searches.

Another way to use usage history is to create *procedural metadata*. Procedural metadata describes all actions and changes that have been applied to the content. For example, typical actions in digital photography include colour corrections, cropping, applying filters like sharpening, etc. All such actions can be stored as procedural metadata, for example, making undoing changes easier. It also makes synchronizing changed content simpler, since it is enough to send only the procedural metadata, i.e., steps needed to recreate the new content from the original content.

All events have some information about them, that is, once again, meta-metadata. They all have timestamps to store the date and time when the event occurred, and one or two *actors*. Actors are used to store the application creating the event, and the person (a contact in the phonebook) that was involved. For example, when sending an e-mail, the messaging application is one actor, and the recipient the other. Obviously, some events only have one actor.

Applications can also subscribe to certain events, or to events concerning certain objects or object types. For example, an application may subscribe to change notifications. When a metadata attribute of the content object is modified, the subscribed applications are informed. This allows a media player application to update its user interface when a song is played, regardless of the application that was used to play the song.

5.4   From Content Management to Metadata Management

We have discussed some of the key issues in content management in a distributed mobile environment, and presented a classification scheme for different types of metadata. We pointed out that mobile personal content is by its very nature distributed across many devices.

Mobile devices are used to create and enjoy content, yet due to limited storage capacity and reduced UI capabilities, content is rarely stored in a mobile device only.

However, a mobile device is well suited for creating content, and accessing it while on the move. It is always with you and always connected. Therefore, with the help of context and metadata, a mobile device has all the potential to become a key device to content management.

Due to the distributed nature of mobile content, the metadata we use to manage all our content must be separated from the content files and binaries. Even when content is stored in various places, in order to manage it we need to know what is available, where it is stored, and how to access it. All that information is metadata separated from the content itself.

We need to start respecting metadata as the first-class citizen that requires its own management tools and processes. Once we separate metadata from the content itself, some of the content management problems become metadata management problems. We need to ensure efficient storage usage, how to keep metadata and its binary content data synchronized, how to handle different versions of metadata, and how to maintain backup of metadata.

Next, we will emphasize the importance of a system-wide metadata management solution over application-specific ones, as a means to solve or (at least mitigate) the issues discussed above. While the reasoning behind this claim is discussed in detail throughout this chapter, in this section we will summarize the key requirements of metadata management framework as:

- Provide a useful service for all content-intensive applications.

- Remove redundancy in both metadata storage and management.

- Aim at run-time efficiency.

- Allow future extensions to content types and metadata definitions.

- Allow interoperability between applications, devices, and content types.

- Support experiences by the means of cross-media relationships.

A requirement for any system is to justify its existence. This means that we just cannot make a theoretically beautiful architecture. It also must work in real life and in all kinds of mobile devices with limited computing power and memory capacity compared to desktop PCs. We claim that by fulfilling the other requirements, this one also is fulfilled.

5.4.1   Cross Media Challenge and Metadata Ownership

Currently, metadata is generally managed by single domain-specific applications, most commonly media players. Music players, such as Apple iTunes, take care of tagging songs with metadata attributes about artists, genre, and so on, but they only care about music-related attributes. Even worse, each music player has its own metadata management functionality that might differ slightly from other players. For example, Apple's popular iTunes media player stores its playlists and song metadata into its own database. No other application can access that data, which results in Windows Media Player or any other MP3 or media player having to store its own metadata into a separate database, duplicating or even tripling exactly the same metadata. Not only are system resources, such as storage memory, wasted but maintaining them are much harder. Similarly, all video-related attributes are managed within video players, and still image metadata in image viewers and editors. Some metadata, including file creation and modification timestamps, are handled by the operating system.

Ultimately, there are lots of metadata that no application cares about. Most applications are unaffected by any metadata beyond the content types they primarily deal with. Some music players can link an MP3 song to an image, such as the album cover art image. The same link the other way round is impossible, as there is no image viewer that plays or even knows songs of the album when you are viewing the album cover image. And which application should be responsible for maintaining the link between songs and album covers?

What is described above is just a simple association between two content objects in two applications. Many times there are associations between many objects used in many applications. The use case with MP3 files received from Eddie during the concert includes links between content objects primarily used in a music player, and in e-mail and calendar applications. Obviously, there cannot be a single application that is responsible for such metadata.

To conclude, metadata is rarely specific to only one application. One of the great benefits of metadata is that it acts as a common language between applications, and therefore facilitates application interoperability. Metadata provides all applications with a means to understand what has happened to any content object before, and shows the big picture (the context) which an individual object belongs to. This also implies that any single application cannot take ownership of metadata and claim that it is solely responsible for it.

Metadata is indeed system-wide information, and its management a part of the operating system. Thus, our framework is designed to be part of the operating system, providing services to all applications with any metadata, for any needs they may have.

Our framework stores all metadata for all content objects only once. This is done regardless of the content type. So music, photos, and video clips are all treated equally. The result is that all applications are provided with a unified access to both content and metadata.

5.4.2 Separating Metadata from Content Binaries

One of the first decisions regarding the metadata framework was to separate metadata from the actual content binaries to increase the scalability and performance of distributed content management. Since metadata is stored in a database with a link to the content binary, finding and updating metadata is fast – there is no need to find, read, and parse metadata from within the content binaries. The framework also indexes the metadata for efficient searching. This means that to find all songs with "Anvil" as an artist, all you need is a simple query to the database, with a list of links to the actual content objects as a result.

There is nothing new in this, since the separation of metadata and content is a common practice amongst media applications. Almost all applications that need to handle large amounts of metadata, such as iTunes or UPnP media servers, do it for the same reasons. The difference in our solution is that all metadata, regardless of the content type, is available for any application.

Storing metadata in a database makes searching faster and simpler, makes it a lot easier to handle metadata that belongs to more than one content binary, and enhances the general metadata management by removing the need to manage the actual binaries.

Having all metadata in a database file makes creating backups easy and fast. Most personal content is priceless, therefore, it is important to keep the amount of content that needs to be backed up as small as possible. Even though editing images, videos, and other media content in mobile devices becomes easier and more popular, as tools and methods eventually improve, modifying the binary content of the media occurs rarely compared to events and other related metadata updates. Every time you play a song or watch a movie, some changes in metadata take place – at least the play count and last accessed timestamps are updated, and perhaps also the ratings or some other metadata is modified. If metadata is embedded inside the content binary, that binary is also changed and flagged as needing backup. Consequently, the whole binary file is transferred in the backup process. When metadata is stored in the database, only the modified metadata rows in a table need to be backed up, making the whole process faster and easier.

There are also some negative consequences to using a database. In particular, separating metadata from the content objects it describes may result in them getting *out of synchronization*. Consider, for instance, transferring the latest hit movie from Eddie's loot to the home PC in order to watch it in your home theatre. What happens to metadata that is stored in the database in your mobile device? Can you move the metadata to the PC, too?

One solution could be embedding whatever metadata there is in the database into the movie file upon transfer. However, some metadata, such as relationships, are hard if not impossible to embed.

Another option is to export the metadata into a separate metadata file (a metadata manifest, see section 5.5), and then also transfer it to a PC along with the content object. In this case, even though all metadata is transferred, relationships to other content objects in the mobile device may still be broken and, obviously, a PC needs to understand the exported metadata format.

Other synchronization problems become evident when you modify some metadata in the PC. For example, the play count of a song should be increased no matter where or how you watch the movie. Another likely scenario is that new metadata is added, and existing metadata updated, due to the easier text input methods that the PC provides. Of course, this new metadata needs to be moved back to the mobile device.

Worse problems arise if the metadata in the mobile device is modified while it is also being altered in the PC. A typical case is listening to the same song in a PC and a mobile device, when the play count needs to be updated both ways. Solving the metadata synchronization is a difficult problem, and there is no automatic solution that is capable of performing flawlessly in all possible cases. There are always situations where the user needs to choose how to merge conflicting metadata or should some metadata be discarded. However, synchronization is such a common problem that many functional solutions exist, for example, OMA SyncML[13] and other protocols that are designed especially for such problems.

---

[13] http://www.openmobilealliance.org/tech/wg_committees/ds.html

5.4.3 Preparing for the Future

Application-specific metadata management has one particular advantage compared to centralized system-level metadata services. Since application-specific metadata management systems are designed solely for the application needs, they automatically take care of any new types of metadata that is needed, and they also support new ways of using this metadata. Dealing with previously unknown metadata on a system level is more challenging.

With a system-wide framework, things are different. No matter how hard we try, we cannot define every possible metadata attribute in the world. Obviously, we can forever try and be proactive, and anticipate all possible new metadata types there will ever exist, and ways to use them, at least until the next software release. However, probably most of these inventions will never be used, just overfilling the framework instead. Therefore, our framework must be designed to be extensible – but without breaking old applications. It is even better if the old applications can use new metadata extensions automatically. Also, the extension process must be as easy as possible, without any central body to solve conflicts like using the same metadata attribute name for different purposes.

Our architecture fulfils these requirements, extending all metadata definitions and their semantics at runtime, without changing the existing applications. Old applications may even, in many cases, use new extensions automatically.

Suppose that an image gallery needs to list all images that the underlying operating system supports, perhaps with metadata fields such as location and title displayed under it. When a new image format (say, a 3D stereo image) is added, the new format will have its own metadata attributes specific to 3D stereo images. The old image gallery using our metadata framework can still find all new stereo images, their location, and title metadata attributes without any changes in code. This is because 3D stereo images are more specialized than regular 2D images, and share many common attributes besides introducing new ones. However, the image gallery does not know anything about the new attributes or what they mean, and probably will not access them, as it is unaware of their existence. How this can be achieved is described later in section 5.7.

Obviously, for every positive feature there is some cost to pay. One of the most significant costs our framework has is due to its nature of being as flexible as possible. In order to support every possible metadata and application, it cannot be optimized for any single application. A single application using its own optimized metadata management routines usually consumes less runtime memory and is faster than using our system built metadata framework. A direct result of per application optimization is that our framework consumes more memory and is slower when you compare it to a single application using its own metadata system, with it using our system wide metadata architecture. So the extra overhead of using our system will never be too much, so that its overall benefits are still greater than any slowness caused by it.

5.5   Overall Architecture

Now it is time to look at the framework that is used to manage all these different kinds of metadata and, consequently, content. In this section, we describe the overall framework architecture, based on a *metadata and content lifecycle*.[14]

Mobile phones used to be referred to as terminals. The word terminal suggests that the phone is the end of the line, where all communications either start or end. Modern mobile phones have now progressed far beyond being mere terminals. Today, the content, especially media content, is also downloaded, created, and especially shared with others using mobile phones – or mobile devices, to be precise. These modern devices are becoming central nodes in content distribution, instead of sticking as end-points. For metadata management, it means that in addition to content, metadata must also travel in and out of the device, which we refer to as the metadata lifecycle.

The framework described herein manages all kinds of metadata from the composer of a song to relationships between content objects. It covers all aspects of the metadata lifecycle. It also provides efficient methods for searching and grouping content based on metadata, in many cases through associations. Figure 5-4 shows generic phases of a simplified content (and metadata) lifecycle, and the required actions related to metadata.

First, in the *Entry/Import* phase, new content enters the system (the "G" phase in GEMS). The content is either created in a mobile device, for instance using a built-in camera, or may have been downloaded off the net, or received as an e-mail. At that point, *harvesters* are set

---

[14] Unlike GEMS, which describes the content lifecycle from the user's perspective, this is a lifecycle from the system point of view. Nevertheless, the phases are similar to GEMS.

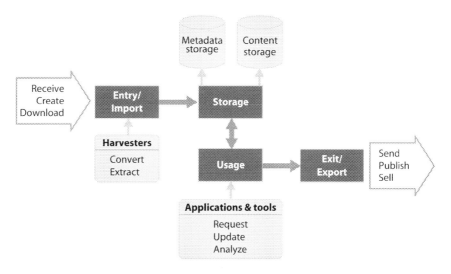

Figure 5-4. Metadata and content lifecycle within a mobile device.

on to the data. Harvesters are components that know the type of the content just entered into the system, and can read and process any metadata the content may already contain. For instance, in the case of downloading MP3 files, the related harvesters know how to extract artist and album information that many MP3 "rippers" put into the file (the ID3 metadata, see section 4.6). Harvesters then convert the extracted metadata into our internal format (described in detail in section 5.6) and store it into the metadata storage, the *Storage* phase. The content object itself is stored separately from the metadata. The use of just one internal metadata format allows managing all metadata in a unified manner, regardless of its original format or usage.

In their normal operations, the applications use metadata in the *Usage* phase (the "E" and "M" phases of GEMS). They request metadata from the metadata storage and show it upon viewing a photo, for instance. They may further search, group, and filter content based on metadata attributes to show, for example, only new songs, or videos shot at a certain location.

In addition to metadata usage, applications also update the metadata or create new metadata that is also stored. For instance, the play count of a song is increased whenever the song has been listened to, and a recipient is associated with a message upon sending it. Furthermore, the content objects are constantly analyzed by an array of software components to identify associations between them, and to recognize objects in a photo, for example.

At the *Exit/Export* phase ("S" in GEMS), when the content leaves the mobile device, *composer* components assume an important role. Like harvesters, these components know what kind of metadata can be included in the content object in question, but while harvesters read metadata, composers write it. Composers read the metadata associated with the exiting content from our metadata database and write into the file whatever metadata can be embedded into it. For example, an ID3 tag composer will write into MP3 file all music related metadata from the database. Similarly, EXIF and IPTC tags are written by respective composers into the JPG files.

Upon exiting, the composers may need to exclude some metadata associated with the outbound content object. A typical reason is that the framework is capable of processing metadata that cannot be included in standard metadata scheme, such as associations in MP3 files. Also, some metadata may have been tagged as private: for example, the user may not wish to expose the shooting location of certain photos. Each metadata attribute has extra information describing whether that metadata can or cannot be exported or sent out of the system.

Of course, our framework can handle a much wider variety of metadata than can be expressed in EXIF or ID3 tags. For instance, as discussed above, all relationships will be lost. In a similar manner, most of the context information does not fit into tags, at least in any standard way. There are a few exceptions, such as EXIF image metadata format (section 4.6), that contains fields for location information in the form of GPS coordinates.

Some content types, such as text notes, will not allow any embedded metadata in the first place. In these cases, our framework can produce a separate file, called *metadata manifest*, containing any metadata that cannot be embedded into the content file. It can also include any relationships that make sense to export. Of course, if you are sending just one photo to your friend, there is not much sense in adding any relationships into the metadata manifest, since the recipient may not have those other items forming the relationship. On the other hand, if you are sending your playlists with the songs, or you have sent other images previously, it might be beneficial to send the relationship information, too. Then the recipient will know how the new content objects relate to the old ones.

There is no widespread standard for metadata manifests. As a consequence, our metadata manifest works only between two devices using our own metadata framework. We hope and believe that there will be a widely accepted standard in coming years. Already there are some initiatives, for example, using Dublin Core in XML[15] or Didl Lite[16] used in UPnP.[17]

There is one more issue worth discussing. Since we emphasize the importance of capturing as much context and usage history as possible, it also means that there are huge amounts of personal and private information stored. Some of that metadata is never intended to be seen by others. For instance, it can be questionable to list of people who were present while a photo was taken. Our framework contains a field for each tag for every object, describing if the tag can be exported or not along with the content binary. This implies that the user can specify that the location for one image is public and fine to export, while exactly the same metadata used in another content object is not exportable. Yes, it is meta-metadata we are talking about.

5.6   Our Metadata Ontology

Although there are various definitions to describe metadata of different content types, none of them fulfils the needs posed by the mobile using environment. Since there is no metadata standard that describes all the metadata types that we need, we have defined our own ontology to describe the metadata and to give a semantic meaning to the metadata. Reasoning and inference tools can then be used to create new metadata that is not self-evident from the metadata properties.

The sections from this point onwards are occasionally technical in nature, and targeted at those readers more proficient in software development. Feel free to skip them if you are not interested in all these technical details, since they will not affect your understanding of the remainder of the book.

The simplest way to store metadata is to store just attribute-value pairs into the database, perhaps using the filename as a key. It does not matter what data is stored or how it is stored; it is just dumped into database. This could eventually lead to many problems, especially since we aim at interoperability between applications. There would be no consistency checks whatsoever. As a result, the length of an audio track, for instance, could be a negative number or even a non-numeric value, since it would be up to the application developer to decide how

---

[15] http://dublincore.org/

[16] Didl Lite schema: http://www.upnp.org/schemas/av/didl-lite-v2-20060531.xsd

[17] http://www.upnp.org

TERM	DEFINITI ON
Instance object	An object created from schema object. It is actual metadata and can be Metadata object, Property, or Relationship.
Schema object	Definition of metadata instance item, i.e., part of ontology and is meta-metadata.
Content object	Correspond to the real world objects, identified by URIs.
Metadata object	Metadata instance object containing properties, relationships, and events. Contains metadata for some Content or Virtual object.
Virtual object	Metadata object that has no corresponding Content object, i.e., it is pure metadata.
Property	Describes characteristics of an object, i.e., metadata tag or attribute.
Event	Describes interactions with content objects.
Relation	Describes dependencies between content objects.
Schema	Describes all definitions in metadata.
Ontology	A class hierarchy which describes each possible metadata object type.

Table 5-1. Definitions for the most important terms used in the rest of this chapter.

to present the metadata. One application developer may measure the length of the song in seconds, another one in minutes and seconds, or a third may make a small programming error and store the file size instead. Interoperability between applications will be quickly lost, since no matter how devotedly we documented the attributes, developers will understand them in a slightly different way or use different attributes to describe the same value. One developer may use kilobits per second as a unit for bitrate, whereas another one could prefer bits per second.

Surprisingly, even metadata standards have ambiguous definitions. For example, EXIF 2.2 defines the artist tag as "This tag records the name of the camera owner, photographer, or image creator."[18] We claim that to solve these problems, we need clearly defined ontologies.

In philosophy, ontology refers to the study of existence. In essence, it aims at answering the question "What really exists?" In computer

---

[18] http://www.exif.org/Exif2-2.PDF

science, and in this book, we define it as "data models that describe all objects that can exist in our framework, including their properties, and the relationships between them." Our whole metadata framework is basically an ontology framework.

What is especially important is that ontology describes and defines accurately all things that can exist in our framework, what kind of properties they may have, how those properties can be used, and what kind of relationships there are between the objects.[19] For example, our ontology declares that a property named *bitrate* exists, and that it can be used as a metadata attribute to describe certain video and music files. It also states that if a bitrate value is used, it must be an integer number with the quantity of bits per second.

Another example is that our ontology declares that there can be only one creation date for any given object, and therefore trying to add another creation date results in an error. Thus, the ontology is used both to ensure interoperability of metadata between applications, and to detect some possible error situations.

Obviously, no ontology can detect all possible error cases. For example, our ontology declares that all songs have a metadata attribute "artist" that can contain any text. If someone enters the name of the composer instead, there is little our framework can do to detect the mistake. Better definition for "artist" and other attributes having a person as a value, would be URIs identifying people. Unfortunately, there are no reasonable solutions available yet, but there are some interesting proposals such as OpenID[20]. The main thing is that semantics and possible values of any given metadata property is documented as accurately as possible and still give application developers and users enough flexibility to cover many possible different use scenarios.

To conclude, without a clearly defined ontology, the system would quickly deteriorate to the situation where each application manages the metadata without any regard to interoperability. We emphasize that all metadata management systems must have clearly defined ontology in order to provide a useful service to more than one application. We also understand the need for flexibility and for creating our own taxonomies (so-called folksonomies) and tags. Our framework

---

[19] Since the ontology describes metadata, it is actually data about metadata, or meta-metadata. Remember we earlier emphasized the importance of meta-metadata? This is because our whole framework is basically about meta-metadata!

[20] http://openid.net

strictly defines a set of metadata schemas, while also allowing creation of new flexible metadata definitions for any purpose.

5.6.1 Instance Metadata and the Schema

Each metadata attribute has semantics associated with it in the form of ontology. Our ontology is not static or something used in designing the framework. It lives in the framework and our framework therefore contains two kinds of metadata. There is the actual metadata describing the content objects, referred to as *instance metadata*, and metadata about metadata, that is, our ontology or *schema*.[21]

All instance metadata – tags, relationships, context, and usage history – is stored in the database as instances of ontology. The framework also stores and manipulates the schema, which allows adding or changing new metadata schemas at runtime.

In theory, the semantics of next to everything can be changed. For instance, a developer can declare that from this point onwards, the attribute *singer* is no longer a valid metadata attribute for songs, but is used to describe Norwegian oil paintings from the 19th century. Obviously, that would result in havoc among all old applications. Therefore, in facing reality, we have limited the ways semantics can be changed. Usually adding new metadata and new meanings is simple and permitted, but modifying existing metadata definitions is not allowed.

5.6.2 Initializing the Framework

When our framework in turned on for the first time, it is completely empty. There is nothing, no ontology, that is to say, the ontology is saying that nothing can exist; it is void. You cannot add any metadata about any object, since the ontology declares that no object can exist, and there are no attributes nor relationships. Therefore, the framework fails the design goal miserably. It is totally useless.

The first thing a developer has to do is to start filling in the void. For this purpose, we have created an ontology that is used to extend the empty ontology when the framework is initialized. Our initial ontology is designed to be useful for mobile personal content, but it can be either totally replaced with other ontologies or, better still, can be extended with new schema objects, properties, and relationships.

---

[21] We occasionally refer to the ontology as a schema since it is schematics of the actual metadata.

Our ontology describes all basic content types found in a mobile device: contact information, calendar entries, messages (SMS, MMS, and e-mail), images, music, video, and so on. In addition to these physical entities, we have different kinds of collections, such as music playlists and photo albums. The collections can be either user-defined or automatic. Typical examples of automatic playlist include recently played songs, and most played songs.

5.6.3 Our Default Ontology

This section describes our initial simple ontology for the first version of our metadata framework. The default ontology creates an initial set of metadata attributes that cover the most fundamental usages for image and music applications. The ontology can be easily extended later by adding new metadata attributes. However, care must be taken when adding new schema classes or changing the schema structure, since they may break the backwards compatibility.

5.6.4 Namespace

All metadata attributes defined here belong to a default namespace. We have reserved the default namespace to be non-extensible to provide the minimum set of metadata definitions that every application developer can trust. All extensions must belong to another namespace, freely selectable by the extension developer. Although our framework only requires that namespaces are globally unique strings and the developer is free to choose whatever means to achieve that requirement, we strongly recommend using URIs as namespaces, as in W3C XML namespace recommendation.[22]

As defined in RFC 3986,[23] URIs are not globally unique strings, so care must be taken when choosing a proper URI to describe a namespace. For example, the URI `http://localhost` is a valid URI according to the RFC, but is not usable as a namespace URI. Any namespace URI must be globally identifying, not just in an end-user context.

The namespace URI does not need to refer to an existing document, but it would be useful for the URI to point to a human-readable document explaining the semantics of the metadata definitions. This would make it easier for other developers to use or even extend those definitions. As an example, a good imaginary namespace URI could be

---

[22] http://www.w3.org/TR/2006/REC-xml-names-20060816/
[23] ftp://ftp.rfc-editor.org/in-notes/rfc3986.txt

```
http://namespaces.example.com/metadata/game-metadata/
```

that would refer to example.com's game-related metadata definitions.

5.6.5 Metadata Schema Objects

This ontology defines five main schema classes that contain all attributes. All instance objects are expected to include at least either the Object schema or the Content Object schema. The Object schema contains the least amount of metadata any object in the system must have. Therefore, the content types of all instances having only the Object schema are basically unknown.

The Content Object schema object adds a few more properties that are applicable to most objects. Using the Content Object schema is strongly recommended as a minimum set of metadata attributes for any object.

In addition to this minimum set of attributes, all instance objects are expected to take one of the more specific set of attributes. For example, all jpg images include the Image schema. Similarly, music content such as MP3s and AAC files need to include the Audio schema and the Time-based schema in addition to the Content Object schema. Videos are an interesting case. They certainly need all Time-based attributes and all Image attributes added to Content Object attributes, but they also may need all audio-related attributes for the audio track.

This ontology has an empty schema object for common media formats. They may contain format-specific metadata attributes but their main reason is to make searching easier. All schema objects can be used as filters, implying that only the instance objects using the schema are selected. For example, with this schema, you can search for all audio objects including either MP3 and aac files, or just MP3 files.

The tables below contain all defined metadata attributes and schema objects. Attribute column gives the name of the attribute, type columns its data type, N column is the *cardinality* (how many time the attribute can be given to a single metadata instance object), and finally a comment giving a brief summary of its purpose. The comment is not intended as a final document for the attribute but is given here only as an example. There must always be a separate document describing the ontology in detail.

A number in the N column means that the attribute must exist exactly as many times as the number given. For example, number one for creationDate attribute says that creationDate must always exist and there must be only one of them in any instance. Numbers may also have a "+" sign, which implies that the attribute can be given multiple times. For example, a value "0+" means that the attribute can be given zero or more times, meaning it is an optional attribute. If an attribute is optional but it exists once and only once, it is indicated with the notation "0,1".

5.6.6 The Most Typical Metadata Schema Objects and Attributes

OBJECT			
Attribute	Type	N	Comment
ID	UUID	1	Unique ID. Provided by the framework.
accessPath	String	1+	Path (URL) to the object. This is not a field but part of the object id.
creationDate	Date	1	Date and time when the object was created.
lastModifiedDate	Date	0,1	Date and time when the object was last modified.
Title	String	1	Human-readable title of the object.
Size	Integer	1	Size of the object in bytes.
Description	String	0+	Freetext description from the object creator.
Freetext	String	0+	This is not an actual metadata field but handled by the freetext indexing engine. Each object may have freetext that is efficiently indexed by the engine, e.g., efficient retrieval
Comment	String	0+	A field for some free-form comments.
CONTENT OBJECT			
Attribute	Type	N	Comment
Rating	Integer	0,1	User-given quality rating of the content object.
Counter	Integer	0,1	Rendering count of the object.
Genre	String	0+	Genre of the object.
Drm	Boolean	0,1	Flag to indicate protected content.
releaseDate	Date	0,1	The date when the content was released to public.
mediaFormat	String	0,1	RFC 2046 media type.

AUDIO			
Attribute	Type	N	Comment
Artist	String	0+	Name of the person performing.
samplingFreq	Integer	0,1	Sampling frequency in Hz.
Album	String	0,1	Name of the album the song belongs to.
Track	Integer	0,1	Track number in the album.
Composer	String	0,1	Name of the composer.
TIME-BASED			
Attribute	Type	N	Comment
Length	Integer	0,1	Length of the object in seconds.
Bitrate	Integer	0,1	Bitrate in bits per second.
IMAGE			
Attribute	Type	N	Comment
Width	Integer	1	Width of the image in pixels.
Height	Integer	1	Height of the image in pixels.
Whitebalance	Integer	0,1	EXIF whitebalance.
zoomLevel	Integer	0,1	EXIF zoom level.
flashUsed	Boolean	0,1	EXIF flag if flash was used.
exposureProgram	Integer	0,1	EXIF exposure program
DVB-H			
Attribute	Type	N	Comment
legalText	String	1	For the TV recordings.

5.6.7 Events

All events and relationships can occur as many time as needed, so there is no N column in this table.

EVENT	COMMENT
Created	Object is created for the first time.
Deleted	Content Object is deleted Metadata object still remains, i.e., framework has metadata about content objects that are no longer available.

Opened	Object is opened for whatever reason. Note, use played or edited events if possible.
Played	Object was opened for rendering but not changed. Metadata change is still possible.
Edited	Object was opened and changed, other than metadata change.
Sent	Object was transmitted from the system by whatever means.
Received	Object was received from outside.
Synchronized	Object has been synchronized with the server. Note the actor field indicates with what server.

5.6.8 Relationships

RELATIONSHIP	LEFT TO RIGHT	RIGHT TO LEFT
Contains	The left object contains the right object.	The right object is a part of the left object.
Replaces	The left object replaces the right object.	The right object is superseded by the left object.
Is version of	The left object is a version of the right object.	The right object is an older version of the left object.
Requires	The left object cannot be viewed/ read/listened, etc., without the right object	The right object is required by the left object.
Uses	The left object uses the right object. This is like Requires relationship but less demanding, i.e., the left object may be enjoyed without the right be enjoyed.	The right object is used by the left object.
References	The left object references to the right object, i.e., it provides information that is based on the right object.	The right object provides base information about the left object.
Summarizes	The left object provides a condensed version of the right object. Can be used, e.g., for thumbnails.	The right object provides more extensive information than the left object.

As mentioned earlier, our relationships are always two-way relationships: they can be read either from left-to-right or from right-to-left. Naturally, the meaning of the relationship changes according to its reading direction. For example, the *contains* relationship, when read from left-to-right, indicates that the first object is a container for the second one. Reading from the opposite direction, the first object is now a part of or belongs to the second object.

Another way to limit the increase of the relationships is closer to a usage guidance than a strict rule. Remember Eddie's loot case? A relationship called "part of Eddie's loot" can be created and then added between every object forming Eddie's loot. Every time a new item arrives, a new relationship between the new item and every old item in Eddie's loot can be added. However, there is an easier and more efficient method, using those virtual objects mentioned in section 5.3. A virtual object called "Eddie's loot" can be created that would have a "contains" relationship between all objects that form the loot. When a new item is received, only one relationship is added (between the loot and the new object). The number of relationships in Eddie's loot would be identical to the number of objects, since each item has exactly one relationship. Of course, items in Eddie's loot may have other relationships with each other, or relationships with objects not part of the loot.

5.6.9 How to Handle Composite Objects

Many, if not most, media files are actually composed of several separate objects. For example, a text document may contain embedded pictures. Sometimes the main object is basically empty (this is the case with all collections) but it still has its own metadata. Our approach to composite objects is to use relationships between objects. In our metadata management system all objects, regardless of whether they are independent media objects such as a jpeg image or a part of another object such as an image in a text file, are handled separately. They have their own metadata independently of other objects, and in the case of a composite object, there is a relationship that indicates whether the object is a part of another object or if it contains sub-objects. See the relationship ontology tables for details about different relationships.

For example, a collection object has a *contains* relationship with all objects that belong to the said collection. Likewise, all objects that belong to the collection have an *is part of* relationship with the collection (or, the *contains* relationship reversed). Naturally, one object may have several relationships with one or many other objects.

Indeed, one of the most used relation in our framework is the *is part of* relation. However, this relation can be used beyond indicating embedded objects or hierarchal relationships between objects. It is also useful in limiting metadata to some fragments of the main object only.

For example, video objects can be considered composite objects containing a series of still images and an audio track (and optionally, subtitles, interactive components, etc.). Using virtual objects that identify a video sequence, metadata properties can be given to any fragment of the video, instead of the hierarchical objects. Then, a whole video has its own metadata properties, certain sequences have their own metadata, and the audio track can have its own properties. For instance, a video clip can be divided into chapters that all have their own titles and perhaps even authors. To achieve this goal, we need to be able to give these chapters, or fragments of the object, unique identifiers. When we can identify the fragment, that fragment may have all metadata attributes that are available to any object.

All that is needed here is to give these individual video fragments a unique URIs to identify them and define a *is part of* relationship with the whole video clip. Using a relation is not strictly needed for storing metadata, but it is useful in that it helps find the parts later by using any of the fragments as a key. Basically, this means that we must have a method to create unique identifiers, URIs, for pointing inside other objects. W3C has defined XPointer to identify fragments of XML documents. Since most objects are not XML, we must define a similar method for these objects, too.

5.6.10 URIs for Fragments

To our framework, all URIs are just unique strings. There is no special meaning in the URIs, and applications are free to use them as they like as long as these strings are unique. However, to ensure interoperability between applications, there must be a common language to the fragments of objects. This is yet another part of the ontology creation.

Our approach is to use existing XML technologies, such as XPath[24] and XPointer,[25] as a basis for a fragment URI. In all cases, for all metadata objects the fragment ("#") part of URI is used to indicate the fragment. This section gives an outline on how the common language might be developed.

For all objects, the fragment ("#") part of the URI is used to indicate the fragment.

5.6.10.1 XPointer for XML Documents

XPointer Framework is a W3C recommendation for fragment identifiers for any resource that has media type of text/xml or one of other

---

[24] http://www.w3.org/TR/xpath
[25] http://www.w3.org/TR/xptr-framework/

xml media types. So it is natural that XPointers are used whenever there is need to give metadata to parts of XML document. Note that the framework does not evaluate whether the pointer is valid. Evaluation is left to the application.

5.6.10.2 Time-based Content Objects

Time-based content objects have duration. Consequently, a point in the object can be simply and uniquely defined by giving a point in time. Typically, video clips and audio streams are time-based content objects.

For time-based fragments, we specify that the start and end times define the fragment. The start time is used inclusively but end time exclusively, i.e., start time is considered as part of the fragment but end time is not. This way, it is possible to create non-overlapping time-based fragments. However, overlapping fragments can be created if needed. As with XPointers, the framework does not evaluate or be affected by how applications use fragments.

Either the start or the end definition can be omitted, but not both. If the start time is omitted, 00:00:00 is assumed and if the end definition is omitted (or if the end definition extends beyond the end of the playback), the end of the playback is assumed.

Time fragment specification is in the form

```
#time(start=hh:mm:ss[:fraction],end=hh:mm:ss[:fraction]).
```

The starting and ending times are expressed by hours, minutes, seconds, and an optional fractional part, indicating how much time has elapsed since the beginning of the playback. The fraction part depends on the type of the media. For video, it is the number of frames from the beginning of the second, and for audio it is milliseconds.

For example, a fragment URI

```
http://path/to/video.mpg#time(start=01:20:00,
end=01:22:00)
```

will point to a two-minute video clip that starts at 01:20:00 and ends at 02:22:00.

5.6.10.3 Coordinate Based Content Objects

Coordinate-based content objects are media objects where fragments can be defined with geometrical shapes that have an area or a volume. Typical coordinate based objects are image files.

The geometrical fragment specification is in the form

```
#img:rect(left,top,right,bottom[,depth])
```

where left, top, right, and bottom define rectangular fragment of the object. Depth can be used for three-dimensional objects. Similarly to time-based fragments, left and top are inclusive, whereas right and bottom are not. This also means that right and bottom can extend over the dimensions of the geometrical object.

Other geometrical shapes can be also used, but only img:rect is currently defined, for example,

```
http://some.server.example/images/ex.jpg#img:
rect(0,0,200,300)
```

will define a top left fragment of image ex.jpg that is 200 pixels wide and 300 pixels tall.

5.6.10.4 Embedded Objects

Objects can be embedded in two ways, by linking or by real embedding. If the objects are embedded using links, the parent object does not contain the other object, but just a link to it. In this case, we do not need fragment URIs to embed the object, since it must have an ID of its own and a relationship can be used instead. Where the embedding is real, such as in the case of an audio track in a video clip, we need fragment URIs.

The fragment URI specification for embedded objects is based on an *object array*. Embedded objects are indexed from the beginning of the document. The first index is 0.

The fragment specification for embedded objects is therefore in the form

```
#object[index]
```

where index is an index to the embedded object, for example,

```
file://docs/word.doc#object[5]
```

refers to the sixth embedded object in word.doc.

5.6.11 Extending the Ontology

As discussed earlier, it is almost impossible to model everything that any software developer may invent in the future. For that reason the ontology is extensible. Software installed on a mobile device can extend the ontology as needed.

We created the default ontology, discussed in this section, to always be available in the framework. The default ontology models all basic data that a modern mobile device handles. This includes photos and images, audio and video clips, and different kinds of collections. More specific types of the known types can be introduced by inheritance. The new metadata definition inherits all properties of the parent meta-data definition. Multiple inheritance is allowed, but the names of the property definitions must not collide; if two classes contain the same property definition, only one of them can be inherited in a subgraph in the ontology.

For example, if a developer needs to create a new image type, a stereoscopic image, they create a new object schema for it. Since a stereoscopic image shares all image-related attributes, such as width and height, the developer will inherit the stereoscopic image schema from the image schema. As a result, all image properties are also now available for the stereoscopic image. Next, all properties that are unique to the special image class can be added. Once the new definition is ready, it can be a parent for even more specific objects. This way, any developer can extend the existing ontology and does not need to create everything from the scratch.

One consequence of the "pick and choose" design is that the ontol-ogy is not a hierarchical tree structure, but an acyclic network of schema definitions. This allows more flexibility in dividing content to different groups. For example, if your application uses a timeline, you may model all content that have time-related properties (instances of such schemas that have chosen time-based schema as their parent). These objects include both videos and music, whereas in a traditional ontology hierarchy, they would be completely separate.

The ontologies are separated with namespaces. Thus, the same class and property names can be used in different ontology extensions without collisions. The default namespace cannot be modified by the developers, but anyone can define an ontology extension and inherit from the default namespace. An ontology extension can be created, and later new schema definitions can be added to the extended schemas, but existing definitions cannot be deleted nor modified. Figure 5-5 shows an example of where a new metadata schema object uses pre-existing definitions.

The system does not only offer basic metadata management ser-vices, but also complex event, relational, and context metadata based queries. The applications can subscribe to receive notifications when a specified item changes, an item of a certain type is inserted, or an event occurs. This allows, for instance, an application to maintain its

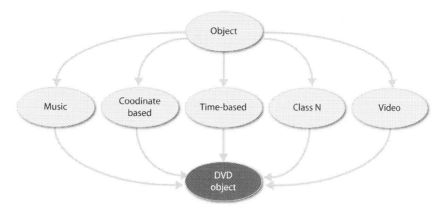

Figure 5-5. An example schema object definition for DVD object.

user interface in an always-up-to-date state without the need to constantly re-execute queries to check for changes.

The ontology is extendible and the extensions as well as the metadata described with the ontologies can be transferred from one device to another. Social sharing, synchronization, and backup features can thus be supported.

5.7   Making a Prototype Implementation

We have implemented a mobile prototype version of our framework on the Symbian platform, commonly found in many current smart phones and multimedia computers. Our implementation uses a SQLite relational database as the underlying persistent metadata store. The object-oriented schemas and the metadata object model are translated into a relational model for persistent storage and SQL queries, but an API abstracts the database structure so that the applications view the data according to the model described in the previous sections.

However, the framework is in no way dependent or even specially designed for Symbian operating system. It has been designed to be implemented on any modern computer platform. The basic requirements for the computing architecture are that there must be a means of creating inter-application services that are available to all applications or processes running on the device, and some kind of permanent storage to keep everything intact across power cycles.

Our design decision to have a dynamic runtime ontology, always available, together with the dynamic extensibility, requires some memory and processing power. It is not intended to be implemented in small memory footprint embedded devices. It should run on mobile

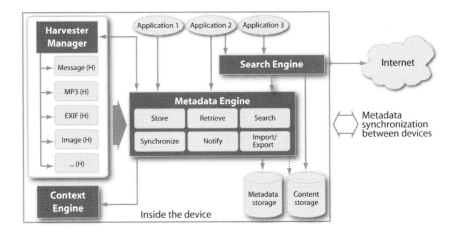

Figure 5-6. Components of the metadata framework.

devices, multimedia computers, and even more efficient devices. There are no problems in implementing it on Microsoft~Windows™ or Linux operating systems.

Our metadata framework consists of four major components that are divided into many subcomponents (Figure 5-6). These four high-level components are:

1. *Harvester Manager* for creating and converting metadata;

2. *Metadata Engine* for manipulating metadata;

3. *Persistent storage* for storing the metadata and actual content; and

4. *Context Engine* to track what is going on in and around the system.

Next, we are going to cover all these components in more detail and take a look at what each component has inside and what its role is in a whole framework. We will not look at their APIs or what functions a developer needs to use those components. Instead, we look at them by using examples to capture their essence and the way they fit into the theme of this book.

5.7.1 Metadata Engine

In GEMS modelling you get the content first, and then interact with it. So it would be rather natural to start the component description from harvesters. However, in the heart of our metadata framework is a component we call *Metadata Engine*. It is easier to understand the

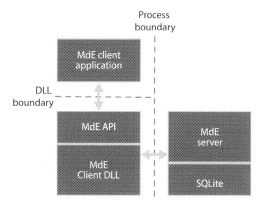

Figure 5-7. Components of S60 Metadata Engine.

other components once you know how we model the metadata and how it is manipulated in the Metadata Engine.

Metadata Engine (MdE) implements client/server architecture where the server stores both the ontology and instances of the ontology (the actual metadata) in its own permanent storage. When a client application connects to the MdE server, all stored schemas (the ontology) is read from the database and transformed from a relational database structure to object-based API classes provided in a DLL. The client application needs to link its code with this DLL. Therefore, MdE provides its clients with the same data model described in previous sections. It also has interfaces for searching, browsing, and manipulating both the ontology and metadata. Figure 5-7 shows the components of MdE for Symbian OS.

The DLL makes all client/server communications and data model transformations opaque to the application developer, so that the developer does not need know what kind of models are used inside the engine. It also allows modifying the models in the engine later if, for instance, we develop more efficient ways to store and manage ontologies. We do hope to be able to use RDF databases natively. However, currently there are no usable mobile implementations available.

Since our first implementation runs on the Symbian operating system, it was natural to choose C++ as the programming language for the first implementation. A problem with C++ in Symbian is that it is difficult to start and experiment with new ideas for applications. SymbianOS C++ is not in any way standard, and there is a high threshold before a developer is able to create anything more complicated than a simple "Hello world" application. For this reason we also decided to create a Python API for developers.

Python is a popular and interpreted object-oriented language that is available for all Series 60 Symbian devices. Python is a powerful language and allows rapid application development in just a few lines of code, while an equivalent application using the Symbian dialect of C++ requires hundreds of lines. Python is an ideal language when trialling and experimenting with anything that is not yet on the shelves. Our Python implementation covers only the API, using exactly the same MdE server as the C++ version. All metadata and schemas written either in Python or C++ are available for applications written in other languages. The MdE server itself is implemented in C++.

The basic unit of our metadata framework is a *metadata object*. Each metadata object must be predefined in the ontology, that is, a schema for the object must exist in the framework ontology. Due to the extensibility, a developer can create a new schema object if necessary, before instantiating the metadata object.

A metadata object can be anything that can be pointed to by an URI. It can be real (a file) or virtual (a collection or playlist). Each object has a set of metadata attributes attached to it. The attributes describe the properties of an object, for example, an MP3 song has attributes such as artist, song title, and duration.

5.7.2 Managing Schemas

When a client application opens a connection to Metadata Engine using the client API, the schema is loaded from the database into the client's memory. The client can add new schema definitions to the database by using add methods available in the API. Deleting definitions from the database is not permitted due to keeping old applications functional by not destroying any functions that they rely on.

The schema is only updated in the client's memory if the client extends the schema itself; modifications done by other clients are not forwarded from the server to the client while the session is open. If it should become important to receive modifications done by other clients, the session must be closed and a new session opened. This is one area where our prototype system needs improvement.

Every definition is defined in a namespace in order to avoid name collisions. The predefined set of definitions is provided in the default namespace. The client can obtain a descriptor containing the URI of the default namespace using the provided method.

It is not possible to add new definitions to the default namespace. Consequently, new definitions have to be made in other namespaces. When a new definition is added, the namespace in which the definition will be included must be specified. The Metadata API creates a namespace automatically, in cases where it did not previously exist.

5.7.2.1 Defining Schemas

The MdE ontology follows principles discussed in previous sections. It consists of metadata object, relation, and event definitions. Now we will take a closer look at these definitions and how they are used in MdE.

A *metadata object definition* is a class with a name and a set of *property definitions*. The metadata attributes of a metadata object are described with property instances. A property instance always relates to a metadata object, i.e., there are no stand-alone properties without metadata object definition.

A property definition has a name, a data type, and a value range. The data type can be any of the following: text, time, Boolean, double, or a 8-, 16-, 32-, or 64-bit signed or unsigned integer. The value range can be used to restrict the possible values of the property instances to a given range. For example, property definition for rating can limit its possible values to a range between 1 and 5.

In MdE, one property definition can be associated with more than one metadata object definition. However, it is recommended to use inheritance instead of giving same property definitions to more than one metadata object class. There are some exceptions, such as when different cardinalities are needed in inherited metadata definitions.

In addition to the type and value, property instances have source and confidence. Source is a URI describing who has created or set the value of the property. Confidence describes how confident the source was about the value when it was set. This is especially required when the property is the result of data mining or some probabilistic algorithm such as trying to detect who is in a photo.

The cardinality of properties is set separately for each class they belong to. The properties can be defined obligatory; also the number of values of a property can be defined. For example, our ontology defines creation date property to be obligatory and unique for all content objects, i.e., its cardinality is one, and only one meaning, therefore every object must have a creation date and no more than one creation date.

The cardinality can also be defined optional and infinite so that the object may have several different values for the same property, or even no value. This is useful in instances of MP3 objects, where some may have several artists and others none. Note that when schema objects inherit properties from other objects, also property definitions are inherited as they stand. There is no way of changing definitions or even the cardinality. However, since the cardinality of the property is set separately to each metadata object definition, it means that it is possible to define one property definition that is optional for some metadata object definitions but compulsory in some other definition.

There is also always a chance to create a new, different definition that has, for instance, the same name but is in a different namespace. The namespace must be different for such re-definitions as it also implies different semantics for the property and its use. As a consequence, from the framework point of view, they will be totally separate definitions.

A *relation definition* is simply the name of the relation. A relation can be described between any two metadata objects, so there is no limit to the class types of the participants in a relation and no means to define such restrictions in the ontology.

Events are almost as simple as relations at the ontology level. The name of an event class describes the event type. In addition, event definitions contain an importance level that can be used to prioritize which events are moved to a secondary storage, should storage space become an issue. Any type of an event can occur to any type of metadata object so, as in relations, there are no restrictions concerning the type in the ontology.

Context metadata is managed by a separate component, Context Engine, in the framework. Details of Context Engine and *context objects* are described below. Context metadata is a metadata object definition where its URI points to a context object stored in the Context Engine, with the possible context metadata attributes defined. For MdE it does not differ from any other content.

In addition to the semantically meaningful properties, keywords can be attached to objects, without defining further meaning for the words. This construct can be used to create a full-text index, for example, of the words in an SMS message.

5.7.2.2 Making Queries

Metadata Engine has three kinds of queries:

1. querying metadata objects

2. querying relationships

3. querying events.

Metadata object queries are used for retrieving metadata objects from the database. The simplest metadata object query is to find all properties of a given metadata object, such as retrieving all properties of a photo. Metadata objects can also be searched based on their properties (find all metadata objects whose 'location' property has the value 'London'). This query would return all objects whether they are images, music, or any other content type. You can also limit the query to concern certain schema or schemas only, for instance, in such a way that only images are returned.

It is also possible to find metadata objects based only on schemas. This is useful if there is a need to list all items that have certain property, such as duration. The search will go through the whole network of definitions, looking for all metadata objects that have the requested property, defined either in the object's schema, or any of its parent schemas.

Metadata Engine supports the standard variety of Boolean operators and conditions on property values. Conditions setting limits to the property values are specified using value ranges. That is, the value must be within the given range in order for the metadata object to appear in the results set. The available operators are:

- Equals(value): a value range that contains only a single value. The property must be equal to the parameter value;

- NotEquals(value): a negated version of Equals(value);

- Less(value): a value range that contains all values less than the given value;

- LessEqual(value): a value range that contains all values less than or equal to the given value;

- Greater(value): a value range that contains all values greater than the given value;

- GreaterEqual(value): a value range that contains all values greater than or equal to the given value;

- Between(min, max): a value range that contains all values between given minimum and maximum values. By default, minimum and maximum are inclusive;

- NotBetween(min, max): Negated version of Between(min, max);

- AscendSection(min, max, isFinal): a value range that contains all values between given minimum and maximum values. The minimum value is inclusive. The maximum value is exclusive if the isFinal parameter is false. AscendSections can be used to cover a larger value range in sections without any overlaps;

- DescendSection(max, min, isFinal): similar to AscendSection, but used for descending values (max > min). DescendSections can be used to cover a larger value range in sections, starting from the highest section;

- Segmented(count, min, max): a value range that divides the range between min and max into a number of equally large segments. The Next method of the Segmented range is used to move to the next segment.

In addition to query conditions, object queries support *property filtering*. Property filters restrict the number of properties fetched from the server, reducing the required data transfer between the server and the client. An application can use property filters to make a query retrieve only those object properties that the application is interested in.

Relation queries are used for retrieving relation instances from the database. An application can query what relationships a given object has or what, if any, relationships two metadata objects have. However, the first implementation does not support recursive querying of semantic network. For example, let us assume that there are three metadata objects with relationships in the Metadata Engine: *song.MP3*, which is part of an *album* that has a cover image *image.jpg*. Our current prototype implementation requires two queries if you are searching for the cover image and you use the song as a starting point. First, you need to find the album the song belongs to, and then, in a subsequent query, to find the cover image associated with the album. A proper RDF query would allow finding the cover using just one query.

Finally, *event queries* are used for retrieving event instances from the database. Event queries have either a time range or a metadata object as parameters. With the time range, you can find all events that have occurred during the given time range. Similarly, with the metadata object you can find all events the given metadata object is involved in.

5.7.3 Why Use SQL and Especially SQLite as Persistent Storage

While designing the framework, we assumed an application developer's point of view. The main focus was to create API classes that match well to our metadata model. The goal was to create an as natural, simple to use, and efficient API as possible. Unfortunately, the amount of complexity in any given system tends to be constant, if not increasing. As a consequence, if the goal is to make the APIs simple for developers, then the amount of complexity increases in the engine part. This is especially true in our framework. We wanted to provide an API that is flexible and useable for all kind of metadata needs. As a result, the management of metadata inside the MdE became more complex.

We basically had three different options when designing our internal data models and how to keep metadata intact:

1. Using files and in-memory data structures

2. RDF databases

3. SQL databases.

Using ordinary files and our own file formats would be the simplest solution. They would incur little overhead, since there are no database engines or other data management systems involved, apart from those we would develop ourselves for internal purposes only. We could relatively easily optimize all data structures to fit our purposes and for the estimated most common metadata queries, rendering the whole design efficient.

However, the optimization is also the weak spot of the file-based solution. The design would be rigid and function as intended only in situations defined at the design stage. The dynamic features were lost for good. However, as emphasized throughout this chapter, one of the most important aspects of the metadata management system is accommodating all kinds of metadata usages, now and in the future. Flexibility is the principal concern for us, effectively ruling out any hand-crafted optimization solutions that restrict the possible use cases, or limit queries to a predefined set.

Another natural approach would be using RDF databases. Our metadata model is close to RDF, and our internal API level data model is indeed RDF, and what we store in our metadata database are RDF triplets.

RDF AND SEMANTIC WEB

Resource Description Framework (RDF) is a language originally intended to support knowledge interchange in the Word Wide Web. Lately, it became a good framework to model knowledge for many kinds of applications. Usually RDF is used with XML but XML is not the only way to represent RDF knowledge. For instance, even though our framework models semantics of metadata in RDF, we do not use XML syntax internally but only when exporting metadata in separate manifests.

In its simplest form, RDF is a triplet, i.e., subject, verb, and object where each part of triplet is identified by URI. Each triplet is a statement about some resource. So with RDF, we can say things like *"Author of 'Alice's Adventures in Wonderland' is Lewis Carroll"* or " 'Alice's Adventures in Wonderland' was released 1865".

Everything in RDF is a resource. It is possible to further describe Lewis Carroll by making statements about him. For example, you can make a statement: "Lewis Carroll is a pseudonym for Charles Lutwidge Dodgson" and "Charles Lutwidge Dodgson was born on January 27, 1832."

Together, all RDF triplets form a directed graph where each resource (i.e., subject, object, or anything that can have a URI) is connected to another (with a verb). Here we make a small extension to pure RDF, since we specify that for every vertex, i.e., link in the graph, there is another vertex in the opposite direction. The result is still a valid RDF, but RDF itself does not make such a requirement.

For more information about RDF, see W3C RDF pages[26] or Berners-Lee et al. (2001).

Indeed, RDF is a tool for creating the Semantic Web. Quoting W3C, "The Semantic Web provides a common framework that allows data to be shared and re-used across application, enterprise, and community boundaries. It is a collaborative effort led by W3C with participation from a large number of researchers and industrial partners. It is based on the RDF, which integrates a variety of applications using XML for syntax and URIs for naming."[27]

That definition is at the heart of what we propose in this book. Our key contribution is not advancing the Semantic Web *per se*, but we are using it as a cornerstone for all mobile content management. We are also extending it and adding components to it to create full management solution for all layers in the stack.

The problem with this option is more practical than theoretical. There are no ready and available RDF databases for mobile environments. At the time of shifting from framework design to implementation, there were several more or less academic test RDF databases available, while none of them were designed to work in mobile devices with limited RAM and stringent power consumption requirements. We are optimistic that future implementations of our framework can also use internally RDF and RDF query languages. One prospective source for mobile RDF query language in the future is Wilbur,[28] an open source project.

```
COMMENT: FOOD FOR THOUGHT

I think it would be important to emphasize the fuzzy
boundaries between "metadata" and "data" (or, to para-
phrase an old saying: "One man's metadata is another man's
data" :-). With the emergence or more and more (seemingly
useful) metadata we find that metadata in itself (separate
from the content it describes) can become useful. And
immediately having made this observation we should
recognize the paramount importance of (meta)data
*interoperability*.

Clearly, there are a large number of different metadata
standards, and through them some degree of interoperabil-
ity is attained. I fear, though, that most of these
standards having been developed for a particular domain
or purpose, interoperability is predominantly achieved
```

---

[26] http://www.w3.org/RDF/
[27] www.w3.org/2001/sw/
[28] http://wilbur-rdf.sourceforge.net/

within these domains. It is important, however, to also achieve "serendipitous" interoperability by being able to link (meta)data from different domains. Here, we should move beyond "rich metadata" to something we might call "connected metadata".

All this being said, we must emphasize the importance of the Semantic Web (Berners-Lee et al. 2001) and associated standards as a promising basis on which such cross-domain interoperability can be built. The Semantic Web represents a novel approach to interoperability: Whereas the "old" approach has been heavy, a priori standardization of *what* is being said in interoperable communication, the Semantic Web emphasizes the standardization of *how* to say these things, and leaves the definition of the semantics of the communication to not require a heavy, committee-based standardization process. We can even envision "agents" that can acquire more information and *capabilities* (= semantic definitions, ontologies, etc.) as they go along and identify the need for such.

Finally, I would like to share my pet metadata scenario: Imagine digital photographs that are subsequently annotated with rich, *connected* metadata (the annotation can in many cases happen more or less automatically: we already annotate images with a timestamp and possibly the location where the image was taken; ultimately the image metadata could include everything we know about the world at the moment the image was taken). Assuming the annotations indicate who the image depicts, the image is now connected to rich descriptions of people – these can come from the user's own address book or may have been acquired from various social network representations (e.g., Friend-of-a-Friend, already a Semantic web-based schema). Assuming that the annotations indicate where the image was taken, a connection is made to any location information we might have (e.g., geographical ontologies and models). Finally, given that the image has a timestamp, it is connected to any other information that has a temporal extent (e.g., the user's calendar). As a result, the metadata acts as "glue" that connects seemingly separate islands of information (people, places, and time). The image itself may now become less important than the metadata that it carries.

Ora Lassila
Research Fellow
Nokia Research Center, Cambridge

(*Ora Lassila is a Research Fellow at the Nokia Research Center, Cambridge and a Visiting Scientist at MIT/CSAIL, working on Semantic Web and its applications to mobile and ubiquitous computing since 1996. Web: http://www.lassila.org/*)

We finally chose a relational database as a persistent storage for metadata objects and schemas and consequently, SQL as a query language. Even though a whole network of relationships cannot be expressed in SQL, it excels in handling simple relationships that form the basis of our relationship model.

SQL has a long history and is theoretically well understood. Furthermore, a countless number of different implementations exist, ranging from embedded devices[29] to huge clusters of data servers. Even though transferring our model into a format that can be stored in a relational table, producing overheads compared to using RDF as a query language, it still seems to be insignificant compared to all positive aspects of using SQL.

There are many different commercial SQL implementations available for the Symbian operating system and other embedded devices.[30,31] We tried several alternatives prior to finally choosing SQLite[32] as an SQL database engine. SQLite is ideal for mobile devices as it is small, self contained (no need for external support components), and is a one file C–library providing almost complete SQL92 query language. It is especially designed to be embedded to other applications and to small devices.

SQLite is an open source project and in the public domain, implying that you can use it for anything you like without any licence limitations or requirements to return modifications back to the community. It also has an active and keen community supporting it. It has been used in various projects all over the world.[33]

Even though SQLite has been ported to almost every existing operating system, there was no publicly available SymbianOS port when we started the implementation. It turned out that porting it to SymbianOS was easy, since it has been written in standard C. And Symbian includes a standard C library for just porting legacy C applications.

In order to flexibly change both the SQL implementation we use, and possibly to change to a totally different model like RDF, we have abstracted the persistent metadata storage using SQLite in two phases. First, for an application developer, using our APIs either from Python or Symbian version of C++ manipulates metadata and ontology purely in our own data model, as described earlier. Before manipulating

---

[29] For instance, http://www.oracle.com/technologies/embedded/index.html
[30] http://developer.sonyericsson.com/site/global/techsupport/tipstrickscode/symbian/p_symbian_0602.jsp
[31] http://www.sybase.com/products/mobilesolutions
[32] http://www.sqlite.org/
[33] http://www.sqlite.org/cvstrac/wiki?p=SqliteUsers

metadata or ontologies, MdE converts data into standard relational model that fits into any standard SQL-compliant database. This is the first phase in converting from our own model into that of SQLite. At the second phase, our SQLite interfacing component takes the standard SQL and converts it to SQLite specific library calls. The reasoning here is that we can also change the persistent storage and the query language in two different places. On the lowest level we can easily change the SQLite into some other SQL database engine. It would require only changes in one component that performs all SQLite specific operations.

More thorough changes will be required if the whole relational model needs to be replaced and, consequently, a query language other than SQL is taken into use. This would require modifying the MdE component that converts our internal model into the relational model. Switching from one relational database engine to another would require some re-writing beyond simple modifications. However, even in this case the API would remain intact.

Figure 5-8 shows the main tables in our relation database and our approach to modelling both metadata objects and ontology using relations. The figure is oversimplified and not used exactly as such in our implementation. Nevertheless, it illustrates the rationale in designing how to convert our internal data model into a relational data model.

5.7.4 Harvester Manager

When new content, whether music, videos, games, or whatever, arrives in a mobile computer, metadata harvesters attack the content to see if they can extract any metadata out of the new content and then transform it to fit into Metadata Engine. Once the new metadata is in the engine, it is available for all applications in the same manner, regardless of its origin or format.

Almost all content objects have embedded metadata inside their binaries. MP3 files usually have ID3 tags embedded by the ripping application, images may have EXIF or IPTC tags created by the camera or image processing application, and there are countless other metadata formats in use. Some harvesters are more advanced than those able to read one format only, and convert it into our internal format. They may look into the content binary itself and understand what features there are in the binary content. For example, some harvesters look into images and try to detect if there are people in them, or even to recognize who a person in the picture is.

The content also has many avenues into the mobile device. It may arrive, for example, in an inserted memory card, via USB, Bluetooth,

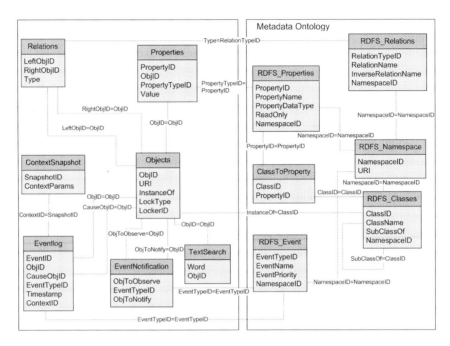

Figure 5-8. Overview of Metadata Engine database schemas.

or as an attachment to an e-mail. In order to catch all incoming content and to choose what kind of harvester is needed, our framework has a Harvester Manager Component that

- recognizes when the content arrives into the device; and

- chooses the correct harvesters to extract metadata out of the content.

Harvester Manager is actually a plug-in framework composed of two kinds of plug-ins. There are *observer* plug-ins that listen to different ports of the device. For example, a messaging observer plug-in listens to all messages that arrive into the device. If a new message arrives, whether it is an e-mail, an MMS, or some other kind of message, the message observer checks if it contains any known media content (for instance, image, music, or video). It notifies the Harvester Manager about the new content, which then chooses a proper media harvester or harvesters to collect the metadata. For example, if the new content is an MP3 file, it launches ID3 tag harvester to extract the ID3 metadata.

We are also designing more advanced harvesters to extract information directly from the binaries. One harvester, still in research, tries to recognize different objects from images, such as faces, airplanes, or birds. Other advanced harvesters may perform signal processing on music to recognize the genre, type, or even artist, even if there is no embedded metadata available. These harvesters usually consume considerable processing power and therefore the battery may soon run out. To avoid these situations, the Harvester Manager also uses context information to launch advanced harvesters while the device is being charged.

Since the Harvester Manager accepts new harvesters as plug-ins, it is possible to dynamically add new harvesters as soon as new content types are born or when new advanced harvesting algorithms are developed.

5.7.5   Context Engine

Context Engine (CE) is a component enabling building context-aware applications by providing a unified access to the context information. Context Engine is a part of our metadata framework, but it is also an independent component that was developed before we began developing the metadata framework. CE runs on a mobile device and handles storing and distributing of all context information in an event-based manner. An application can subscribe to the context information, and eventually receive notifications from the CE whenever there are changes in the environment (Figure 5-9).

Every application can either produce or consume context information. Context producers provide new context information, such as location and nearby Bluetooth devices, to the CE, while the context

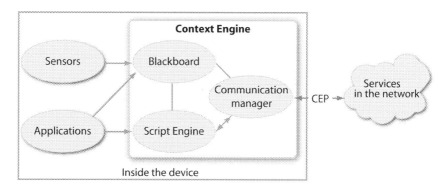

Figure 5-9. The components of Context Engine.

consumers use this information to enable context-awareness. Of course, most applications are actually both context producers and consumers simultaneously.

Context Engine uses *blackboard architecture* to share context information. In the blackboard approach, a memory area acts as a central repository for shared data. The blackboard is equally available to all applications and sensors. Sensors usually put data onto the blackboard, but other applications can check the blackboard for current contents. Such applications can then apply specialized knowledge of some particular contexts based on that data, and finally put new further processed data back onto the blackboard, making it effectively available to other applications.

Context information covers a broad range of different data types. Very low-level sensor information can be as relevant as more abstract information, such as "The sun is shining." The CE can store all kinds of data, making no difference between data types.

Like metadata, contexts also need ontology to allow interoperability between context-aware applications (Figure 5-10). For most context-aware applications, it is enough to define application-dependent meaning or semantics. However, we have defined ontologies for some common context data types, too. The context ontology is simpler than metadata ontology, since we do not need to include relationships or usage history, nor property inheritance for context data. However, the context ontology must also be extensible for the same reasons as metadata ontology. We cannot know all possible context types that exist.

Context information is represented in the Context Engine as *context objects*. Context objects encapsulate context data and context meta-

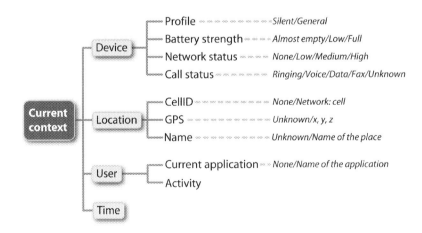

Figure 5-10. A part of Context Ontology.

data. Context data has the current value of the context, and the meta-data describes the context. A context object can store the current temperature, for instance, and the metadata states from where the data was obtained, when it was last updated, and how accurate the context information is.

Context Engine can have one or more *Script Engines* that can make decisions based on context information to create new more meaningful context information. For example, if the temperature is +25°C and the luminance value is high, we can conclude that it is a warm day. This reduces the traffic between the applications and the CE, as applications do not need to query each context data separately. Furthermore, some applications need several context objects to be at certain values before they react. If all of the conditions can be checked in the engine, the separate context values do not need to be sent, only the resulting abstract context.

Not all context information is created inside the terminal. For instance, weather, whereabouts of friends, or even location, can be provided by external sources. All external context sources use Context Exchange Protocol (CEP) to exchange context related information with Context Engine. CEP is used in open source Multiuser Publishing Environment tool and the specification of CEP is freely available from MUPE site.[34]

5.7.5.1 Context Terminology

Unfortunately, context awareness is still a relatively new area. As a result, there are not yet clear and fixed meanings for all terms. Therefore, the definitions here are not necessarily the same as elsewhere, although we keep the terms fairly generic.

Context Object is a basic unit of context information. It has a semantic meaning defined by some ontology and a range of values. Context Engine itself does not care about semantic meanings, but obviously applications using context objects attach meaning to it.

All Context Objects in the Context Engine have a *type* (or sometimes a name is used instead of the type) and a *source*. As far as CE is concerned, both are plain strings that uniquely identify the object. Types are defined in context ontology, thereby declaring the semantic meaning of the object. Context Engine treats the type as a path from the root to leaf node in a tree that is defined by the ontology. Each arc between nodes is separated by a ":" sign. Subscriptions to types are considered as path prefixes, so events are generated for all types

---

[34] http://www.mupe.net

that start with the same string as given in the subscription. As an example, if an application subscribes to "Location" context, any change in "Location:GPS", "Location:CellID", and "Location:Name" context objects will be sent to the application.

The source uniquely identifies the creator of the context object (the context producer) who created the context object or changed its value. We recommend that a URI is used as the source string. One provider can produce different types of context objects and any given type of context object can be produced by several producers, but the type and source together uniquely identify any context object.

Each context object also has a value. Context Engine handles all values as strings but, again, the real interpretation of the value depends of the semantics of the type.

Context Engine maintains some metadata with every context object. Every object has a timestamp that is updated every time the value of the context object is changed. Applications can also add a confidence value to any context object they create or change. Also, CE maintains a history buffer for every context object, so old context objects can also be retrieved. However, there is no guarantee on how many old objects are stored. For historical reasons, CE takes care of its own metadata, even though its metadata should be in Metadata Engine.

5.7.5.2 Scripts

Scripts are small "context-aware applications" that run on behalf of their owning application in the Script Engine component of Context Engine. Each script has a condition that is related to values of one or more context objects. When a condition becomes true, one or more actions are carried out. Actions can include inserting new context objects into the blackboard where all context objects are stored, or sending a notification message to the owner application.

There is no restriction on what kind of applications can use scripts. To external applications (those running outside of the mobile device), the scripts are the only way to get access to context objects on the blackboard. Script Engine provides more complex subscription facility to applications than the blackboard.

Scripts can be sent over the air in XML format by using facilities in Communications Manager. The script XML specification is available at the MUPE homepage.

Applications have a possibility to delegate subscriptions to context objects to Script Engine. If the purpose of an application is to create higher-level context objects, it might be easier to use the script engine rather than Context Engine directly. Of course, as discussed above, for those applications that run on a network outside of the mobile device,

Script Engine is the only way to access the contexts in Context Engine.

Using scripts is simple, all scripts having one *condition* and one or more *actions*. When the condition becomes true, the actions are carried out. The current version of Script Engine can perform two actions, "set atom" and "notify". The *set atom* action creates a new context object that is put onto the blackboard. The *notify* action creates an event with a message that is sent to the application that owns the script. It is the duty of the application to process the message. For example, Context Engine's Communications Manager understands messages that are SMS URLs. One script may contain both types of actions.

5.7.5.3 Communications Manager

Communication Manager (CM) is the component in Context Engine that is in charge of all communications with external applications and context sources. It is analogous to harvesters and composers in the overall metadata framework. CM can receive requests for reading and writing context objects from external sources using Context Exchange Protocol. For example, if an external application needs to know the location of the user, it can send the query using CEP to Communications Manager. CM first checks if the external party has rights to read the data and if it is allowed, reads the location context object from the blackboard and finally sends it back to the requester. External parties can also send scripts to CE. Whenever this kind of script is fired, that is, it becomes true, the CM is notified by the Script Engine. It is then up to Communications Manager to notify the external party if considered necessary.

5.8   Facing Real Life

As a system component, the framework incurs a performance penalty to all applications using it. Since it is intended to manage all content in the mobile device, almost all applications are affected. Obviously, the applications that manage large amounts of content, such as image galleries, phonebooks, and music players, are heavily affected. Unfortunately, those are also among the most used applications for many mobile device owners. What follows is that the framework must be implemented carefully so that it performs optimally, by making compromises where they are needed yet still providing the best possible service.

One of the most basic choices in computer programming is choosing the balance between memory and speed. Simplified, this stems from the fact that storage memory is cheap, slow, and usually there is plenty of it, while execution memory, RAM, is expensive, fast, and a limited resource. Therefore, the programmer may choose either to consume more RAM to increase the execution speed, or decide that conserving it is more critical and design data structures that are mostly stored in the storage memory and read into RAM only when needed. Of course, reading data frequently from slow storage memory to RAM slows down the execution time significantly.

5.8.1 Memory Consumption

Our Metadata Framework reads the whole schema into the application's memory space upon connecting to the MdE server. It means that some RAM is consumed whenever a new client connects to the engine. We decided that providing runtime integrity checks and enforcing that metadata is always used as intended is so important that the usage is justified. This is especially since the schemas usually have a small memory footprint. One object definition usually takes only hundreds of bytes of memory, and the whole ontology takes a couple of kilobytes. With current devices equipped with megabytes of RAM, this is acceptable.

Runtime memory consumption of metadata objects is another issue to consider. In the worst case, creating a huge metadata object with thousands of properties can consume all available memory. In most cases, where applications do not intentionally try to trash the memory, the use of memory is not an issue. However, there are a couple of exceptions, such a Digital TV case described below.

The amount of metadata for basic content types such as images, songs, and other documents, is a few properties per items, usually no more than a couple of dozen. The amount of required memory may be as high as a few kilobytes, but usually metadata resides in the database and is used only in small chunks at a time.

Digital TV content produces some challenges. The DVB standard used in Europe and elsewhere usually have program guide information for a week. The total amount of EPG data can grow to hundreds of megabytes. Of course, all of it is not needed at one time, but it still needs to be stored in the database and also harvested on arrival. DVB EPG is not yet widely used in mobile devices, but we need to proactively design how to process, store, and use megabytes of data in the framework. The current system is not yet optimized for such a usage.

5.8.2 Speed

In addition to memory consumption, another performance issue is execution speed. Again, the flexibility of the system, and the way it enforces that all metadata properties and objects adhere to some ontology, causes performance penalties in execution speed compared to a simpler application-dependent metadata management system. There is no way around it; our framework is inherently slower than any optimized metadata management system for a single application. We can increase the query speed and the overall performance by using in-memory caches and indexing the data in all possible ways, but those consume more memory. The main problem is choosing how much memory we can use in order to gain speed or how slow the system can be while still remaining usable.

Practical tests with our non-optimized framework show that reading or querying metadata from Metadata Engine is not an issue. However, writing data into the system tends to be a problem.

Metadata Engine is designed to be used in environments where new metadata objects are mostly user-created or come via e-mail, MMS, or by using other messaging systems. There are no significant issues as long as creating a new metadata object and adding its properties only takes a few milliseconds. The real challenges are related to importing tens of thousands of media items in one go. This may occur, for example, when using our framework for the first time and the user wants to transfer all metadata from an old media server to our framework. It will take time.

What we can learn from the above is that the system must remain responsive, even while importing large amounts of metadata. Therefore, our APIs are required to be asynchronous whenever there is a delay above 500 ms (such as while importing tens of thousands of images). Asynchronous API calls carry out the operation in the background, allowing the user to perform other tasks simultaneously.

Importing metadata usually needs to be done only once. However, our framework also regards all user actions as metadata and stores them too. All human-initiated events, including opening a media item, watching it, and editing it, occur somewhat infrequently from the computer's point of view. No one can create usage events rapidly enough to impose any effect on the system. Applications, however, can do it more easily. When designing an application that constantly adds new metadata objects, modifies them, and thus creates a torrent of events, then our framework is not best suited to that task.

The speed of the system is one of the major issues of any system component. They must be optimized for both their memory consumption and execution speed, since no matter how good a service they provide, if they make applications unusable, they are futile.

While the development is continuing, the initial results are highly promising. The performance of the metadata system appears to be satisfactory in terms of access speed and memory consumption, considering that we are using a non-optimized build. We have tested the system with collections of mobile media files, where the system performs as expected.

5.8.3   Example Usage of Metadata Engine

Let us take a short walkthrough with Metadata Engine, and see how a developer (we shall call her Faye) can make use of it in content management. We assume that the ontology presented in our Metadata Ontology already exists in the engine.

Faye is developing an application that can play MP3 music using playlists, regardless of their storage location (songs can be stored either locally, or remotely on a network server).

First, she notices that there is no MP3 schema in the MdE. This is not the case in the real version, but let us assume so in this example. It means that she will have to create a schema definition for MP3 files before any MP3 specific metadata can be used. She decides to extend the already existing music definitions so she uses the API methods to get the current definition for music object, and then with another method she creates a new MP3 schema definition extending the old music definition. She uses her own namespace to avoid property name collisions with any other schema.

Faye then creates ID3v2 property definitions for all attributes in ID3v2 that are not already included in old music definition. As a side benefit for creating MP3 schema object, she can use MdE as an MP3 search engine to find all MP3s easily and quickly by using it as an MP3 filter.

Faye's application scans all MP3 files in the device – she creates an MP3 harvester. Besides scanning the device itself, she makes the harvester scan all available songs from a DLNA certified media server in the home network. She makes it create an instance of MP3 schema object for each MP3 file it finds, thus creating a metadata instance for each of them. The harvester adds all ID3 tags it finds inside the MP3 file or from the media server as metadata properties into the instance object. Instance objects contain actual metadata which is stored into database inside MdE as MP3 metadata. Again, our system harvesters will do all MP3 metadata importing automatically, so in real life there is no need to make another one unless there is some special metadata that needs to be harvested.

Once the harvester is ready, the metadata it creates is available for all applications, not just for Faye's. Those programmers who know Faye's namespace and the semantics of her metadata can use all of the metadata her harvester creates, while others can still use those attributes that were already defined in the music schema.

Faye's MP3 application also has playlist management functions; thus, it needs metadata about playlists. Similarly to MP3 metadata, she creates a schema object *MP3_playlist* by extending an existing playlist object. She gives it one new Boolean property definition, *random play*. The property definition she creates dictates that the property can exist only once for any MP3_playlist, and if it exists, is must have a single Boolean value.

Faye states in her metadata namespace documentation that if the value of random play property is *true*, then the playlist must be played in random order, while if the value is *false* or non-existing, the songs are played in a linear order. Now if a user of her application creates playlist, the application creates an instance of an MP3 playlist schema object. One such instance may have a URI `playlist://favorites` with a title property (derived from the parent definition of MP3_playlist) "My Greatest Hits" and random play set to true.

When the end-user of Faye's application will add songs into the playlists, the application will create a new *contains* relationship between the playlist and the song. Once the application has created a *contains* relationship from the playlist to the song, the MdE will automatically create a new *is part of* relationships from the songs to the playlist.

Perhaps one of the songs will be downloaded from some Internet music store and has been protected by DRM. Therefore, its metadata instance also has a DRM flag property set to true. For these situations, Faye programs her music application to add *requires* – relationship from the song to the DRM certificate it needs. And again the MdE automatically creates relationship *is required by* from the DRM object to song. Faye can now make her application show immediately if some MP3 song cannot be played since it misses the DRM certificate.

When Faye runs her MP3 player, it quickly obtains a list all MP3 songs in the device, since it simply requests all metadata objects that are an instance of the MP3 schema. For MdE, it is a simple database query. The application automatically obtains all MP3 songs stored in her media server, since the metadata about them is also stored in the device. The application can also display all basic information about them quickly, since the metadata is immediately available and there is no need to read binary files.

Each time the song is played, Faye programmes her application to send an event to be stored as metadata in the MdE. The stored event includes the ID of metadata instance of the played song, the timestamp when it was played, and IDs of nearby Bluetooth devices.

Faye makes her application create automatic playlists, again by creating instances of MP3_playlist schemas, by using the event data. She creates a *most played songs* playlist where relationships between the playlist and songs are automatically updated based on how many times the song is played. She also makes her application to create *friend* playlists for each of Bluetooth ID discovered while a song is played. Each time a song is played, while a friend, or more accurately a friend's Bluetooth device is nearby, her application creates a new relationship between the song and the friend's playlist associated to that Bluetooth ID.

5.9   Metadata Processors

Our metadata architecture allows and encourages collecting personal and private data as metadata. This includes context information, usage history, logs, and lots of other information that is useful for providing personalized and easy-to-use services. On the other hand, this data must be guarded carefully and should never be revealed to untrustworthy parties without the user's permission. Our architecture provides an object-level access control to any metadata items. This means, for example, that the user may allow location information to be revealed in some pictures but not in others. Furthermore, SymbianOS version 9 has a built-in platform security that mandates certificates for all applications, effectively providing a way to restrict access to metadata architecture for only trusted applications. One means of dealing with privacy is *metadata processors*, small applications that can be used for both privacy control and management of new and yet unknown metadata.

A metadata processor is simply a DLL that is associated with a metadata property. That is, there is a relationship with the metadata property and the DLL. Whenever the property in question is either read or written, the DLL is called with the data to be modified, and the ID of the calling application as parameters. SymbianOS version 9 requires unique IDs and certifications for all applications, so the processor is guaranteed to know, or not to know, what application is actually requesting the data or trying to modify it. The processor may then either accept the new data as it is from the application, or refuse it, depending on whether the application is trusted or not. Similarly, the processor may refuse to give the data to some applications it does not trust.

Additionally, as well as accepting or refusing, the processor can modify the data. For instance, it can blur the property data before it is delivered to the application from the Metadata Engine. Blurring can be used to make some private data less private. For example, instead of giving the exact location where the picture was shot, only the country is returned to the requesting application. Similarly, instead of giving out names of the participants in a calendar invitation, only the number of participants is given out.

Processors need to be written by those who create their own metadata definitions. The framework will not associate a processor to any metadata property, but only to those that belong to a namespace that is owned by the processor writer. Yet again, this requires that the underlying operating system can enforce authentication of processor DLLs and every use of the Metadata Framework API calls.

Another usage for the processor, beyond privacy control, is assist old applications to use new metadata definitions. Applications developed prior to the releases of some new metadata definitions cannot use the new definitions, as they cannot know what they mean or how to use them. Since processors are basically just applications and their writers know the purpose of the new definitions, the processors may, for instance, use its own dialogues, displays, etc. to show meaning of the metadata if it makes sense. A simple example would be if someone creates a new type of location metadata. If an application has no idea how to use the new location definition, the processor attached to it might show the location using its own maps or other mapping applications.

5.10 Summary

In the previous chapters, we discussed how metadata is essential for content management. In order to solve problems in content management, and especially those that distributed nature of mobile content brings, we also need to learn how to manage metadata.

We showed the variety of the metadata ranging from simple tags to context and relationships. We discussed how it is important to make the metadata management framework something that the operating system takes care of, instead of the applications themselves, and why it is important to separate metadata from the content binaries. We described our extensible ontology and discussed how it is important to define semantics of metadata as accurately and strictly as possible for ensuring application interoperability, but still support easy and flexible end user metadata tagging. We also discussed how it is important to create future-proof ontologies, since we cannot know all possible metadata needs now. There will always be many content types in use and new ones will be invented to solve yet unknown needs.

However, we cannot create yet another metadata format but our metadata model must work with all current and future formats. Thus, we only created an internal API that applications can use, not the format itself. We incorporated harvesters into our metadata model to convert metadata from existing formats into our internal format, and vice versa.

Finally, we described our prototype framework that we have implemented on Symbian S60 devices. We went through its components and their functions and also how it is intended to be used.

5.11 References

Adjeroh D.A. and Nwosu K.C. (1997) Multimedia Database Management – Requirements and Issues, IEEE MultiMedia, Volume 4, Number 3, pp. 24–33, July 1997.

Berners-Lee T., Hendler J., and Lassila O. (2001). *The Semantic Web*. Scientific American Volume 284, Number 5, pp. 34–43. Available online at: http://www.sciam.com/article.cfm?articleID=00048144-10D2-1C70-84A9809EC588EF21.

Bernstein P.A., Hadzilacos V., and Goodman N. (1987) Concurrency control and recovery in database systems. Addison-Wesley Longman Publishing Co., Inc. Boston, MA, USA.

Boiko B. (2005) Content Management Bible (2nd edn.). Wiley Publishing, Inc. Indianapolis, IN, USA.

Dingledine R., Freedman M.J., and Molnar D. (2001) Free Haven. In: Oram A. (ed.) Peer-to-peer: Harnessing the Power of Disruptive Technologies. O'Reilly & Associates, USA, pp. 159–187.

Hirata K. and Kato T. (1992) Query by Visual Example – Content based Image Retrieval. Proceedings of the 3rd International Conference on Extending Database Technology: Advances in Database Technology table of contents, pp. 56–71, Springer-Verlag, London, UK.

Katz R.H. (2002) Adaptation and mobility in wireless information systems. IEEE Communications Magazine, May 2002, Volume 40, Issue 5, pp. 102–114.

Lin F. (2002) Multimedia and multi-stream synchronization. In *Distributed multimedia databases: techniques & applications*, pp. 246–261, Idea Group Publishing, Hershey, PA, USA.

Marpe D., Wiegand T., and Sullivan G.J. (2006) The H.264 / MPEG4 Advanced Video Coding Standard and its Applications, IEEE Communications Magazine, Volume 44, Number 8, pp. 134–144, August 2006.

Mills D., Levine J., and Schmidt R. (2004) Coping with Overload on the Network Time Protocol Public Servers. 36th *Annual Precise Time and Time Interval (PTTI) Systems and Applications Meeting* (PTTI 2004), December, 2004.

Mohan R., Smith J.R., and Chung-Sheng Li (1999) Adapting multimedia Internet content for universal access. IEEE Transactions on Multimedia, March 1999, Volume 1, Issue 1, pp. 104–114.

Raymond D. (1998) Searching for facts in all the wrong places. *Machine Design International*, Volume70, Number 5, 1998, pp. 51–57.

Salminen I., Lehikoinen J., and Huuskonen P. (2005) Developing an Extensible Mobile Metadata Ontology. *Proceedings of the Ninth IASTED International Conference on Software Engineering and Applications*, pp. 266–272. ACTA Press.

Tseng Y-Y. (1999) Content-based retrieval for music collections. Proceedings of the 22nd annual international ACM SIGIR conference on Research and development in information retrieval, pp. 176–182, ACM Press, New York, NY, USA.

Chapter 6: User Interfaces for Mobile Media

In the previous chapters, we provided an insight into personal content experience. This included the characteristics of mobility and personal content, the importance of metadata, and the design of a software framework that enables an enjoyable personal content experience through efficient management and analysis of content, context, and metadata. Now it is time to turn our focus to the end-user's perspective.

Shifting the focus means examining what the user wishes to accomplish (the tasks) and the parts of the device that are available for performing those tasks. The GEMS model, described in section 3.6, can be used as a basis when inspecting the relevant tasks related to personal content experience.

The personal content management system, including databases, Metadata Engine, and other components described in Chapter 5, are

Personal Content Experience: Managing Digital Life in the Mobile Age   J. Lehikoinen,
A. Aaltonen, P. Huuskonen and I. Salminen © 2007 John Wiley & Sons, Ltd

hidden from view. A layman is usually not interested in the technology that is involved in the operation of the personal content management system, taking them for granted. The usability and usefulness of the whole system is judged by the user based on the user interface, that is, the components of the system that the user interacts with. Therefore, the role of the user interface is not to be underestimated!

There are numerous devices that are considered *mobile*, based on their battery life and the tasks they are designed for (see Chapter 2 for discussion on mobility and the classification of mobile devices). Also, a mobile device should be small, lightweight, and fit into the user's hand or pocket, being easy to carry at all times. Such a device may also be worn in an arm or wrist band or carried on a neck strap or a belt holder.

We concentrate on devices suitable for *continuous* and *nomadic* mobile use (Rosenberg 1998). The use of a device is considered continuous if it can be operated while the user is actively moving (walking, running, or riding a bike, for instance). However, operating the device is not usually the primary task. On the contrary, in nomadic use situations, the user may be temporarily stationary, such as sitting in a cafeteria or standing at a bus stop. Using the device while sitting in a moving vehicle can also be considered nomadic. A common characteristic for devices in both cases is carryability, due to the demands of mobility.

Mobile device manufacturers are paying considerable attention to the device's ease-of-use. Nevertheless, using them is becoming increasingly difficult due to the *feature race* (the growing number of features), driven by two major trends:

1. *user pull*, which means that people want more versatile gadgets, and

2. *technology push*, which implies that the device manufacturers add new technology that is regarded as a user benefit, or a selling point or differentiation point from the competitors.

The majority of mobile device users claim that the devices have become more complex and difficult to use, yet the hard evidence is difficult to find as a *feature* is ill-defined at best. However, studying the development of the number of text strings displayable to the user in mobile phones reveals some trends. In 1992, the first-ever GSM phone Nokia 1011 contained 406 display texts, but 10 years later the Nokia 6610 had 3085 texts strings (Lindholm et al. 2003). In essence, the number of strings (not words or characters!) is 7.5 times higher, even though the user interface has not changed dramatically.

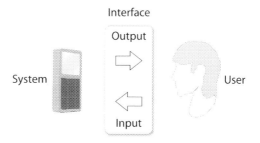

Figure 6-1. The main components of the interactive system.

A mobile device can be considered an interactive system that aids the user in accomplishing their goals. The term *interaction* refers to communication between the system and the user through an interface, composed of channels for *input* and *output* (Figure 6-1), usually referred as the *user interface*. Interaction is based on different human communication channels that we discuss briefly in section 6.1 (Dix et al. 2003). This chapter discusses some general characteristics of interactive systems, focusing on mobile aspects of user interfaces targeted at personal content experience. We employ a number of in-house UI design concepts to illustrate the topics discussed.

Each component of an interactive system corresponds to a translation from a component to another during one interactive cycle. The cycle starts in execution phase, when the user formulates a task for achieving a goal and starts manipulating the system through input. As a result, the system transforms itself, based on the operations translated from the input. The system is in a new state that needs to be communicated to the user during the evaluation phase. Therefore, the current values of system attributes are presented to the user. Then, it is the user's task to observe the output and evaluate the results of interaction in relation to the original goal (Dix et al. 2003). Nonetheless, users do not live in a vacuum: a mobile device is operated in a certain context that affects its use to a great extent.

Prior to discussing users, interactions with personal content, and mobile user interfaces, we will briefly review the evolution of the user interfaces for computing devices.

In the early 1950s, at the dawn of the computing era, use was based on punch cards that were fed into readers, the output then printed on paper. As such, the user could not interact with the system directly, which hindered its use in many ways. The beginning of the 1960s introduced command language user interfaces, as first keyboards and alphanumeric displays became available. This era bloomed until the mid-1980s, when the third UI generation made a commercial break-through, led by Apple Macintosh and later by Microsoft Windows. This generation is called WIMP GUIs (Graphical User Interfaces based on Windows, Icons, Menus, and Pointing devices). The majority of us are still using it, though its deployment started over 25 years ago (van Dam 1997).

The work for the next generation (non-WIMP) UIs, which aim at using natural ways of interaction via, for example, speech, gaze, and gestures, has been ongoing since the 1990s, but they have not yet made a commercial breakthrough (van Dam 1997).

Even though we focus on the mobile descendants of third-generation user interfaces, which rely on graphical presentation of information and visual sense for acquiring it, we do not wish to under-estimate the importance of other types of interfaces, such as speech user interfaces (SUI). It is tempting to consider genuine eyes- and hands-free use of mobile devices with the aid of speech recognition and synthesis.

6.1   Human in the Loop

Every person is unique, but since we belong to the same species, we share common physiological and physical characteristics, regardless of our ethnic origin, education, background, culture, and so forth. Because the systems that we design are targeted at humans, we briefly introduce some of the general characteristics that affect the user interface design, regardless of the application and the user segment. For more infor-mation, see Carrol (2003); Dix et al. (2003); and Norman (1990). Understanding the characteristics of a human is important from the design point of view, since they can be translated to general UI design principles (section 6.3).

When studying a human as a part of the interactive system, we must first consider how we work together with the interface and what our main channels for input and output are. With humans, the input relates to perceiving the world with the senses of sight, touch, and hearing, in order to receive information and make decisions about our next actions. In most interfaces, vision and hearing play central roles in receiving information; vision has a higher information bandwidth com-pared to other senses, followed by hearing.

Vision has a number of physiological and cognitive aspects that affect perception. For instance, there are limits on how small objects can be or how many colours a human can perceive. Furthermore, due to cognitive visual processing, there are certain ways (Gestalt laws) that determine how people group objects based on proximity, similarity, symmetry, and so on. A special case in visual perception is reading, a combination of visual and cognitive processing.

Similar to vision, the human ear can sense frequencies from 20 Hz to 20 000 Hz, being most sensitive between 500 and 2000 Hz, the frequency of normal speech.

The human output relates mainly to motor control of effectors, such as fingers, limbs, and the vocal system, and depends heavily on physical characteristics, skill, and practice of the user. The motor control can be measured, based on reaction and movement time.

Reaction time depends on the senses. For example, the user can react to an auditory signal in 150 ms, but reacting to visual signal requires 200 ms. The second measure can be estimated by using Fitts' law (Fitts 1954).[1] Processing the received information and translating it to motor control requires various actions related to, for example, problem solving, reasoning, and memory.

Humans have a limited capacity of storing information. First, a stimulus received through the senses is buffered into sensory memory for 0.5 seconds. It is constantly overwritten by new information and only a small portion of it is transferred into working memory. Second, the size of working (or short-term) memory is limited to 7 ± 2 chunks of information (Miller 1956). Consequently, from the user interface design point of view, it is imperative to minimize the amount of information that the user needs to remember, and primarily present information that is relevant to the current task.

All the things that we "know" are stored in long-term memory; transferring information from working memory to long-term memory requires learning. Naturally, the frequency of use and time required to learn affect this. Since systems are used by people that are at various stages in the learning process, all of them should be supported. Typically, novice users require more navigation steps and aid when performing a task, which may be straightforward to complete by expert users (Dix et al. 2003; Shneiderman 1998).

Related to problem solving, a central concept is a *mental model*, a representation of reality for understanding some phenomena based on prior knowledge. These models help to understand experiences, predict the outcomes of actions, and handle unexpected occurrences (Norman 1990).

---

[1] Fitts' law is used to predict *movement time* from a starting point to a target region with the aid of empirically determined input device -dependent constants.

6.1.1 Searching

Searching is an important and complex cognitive process that is often related to personal content experiencing. It is often required to locate a piece of content, and several technical aspects are required to achieve efficient searching.

Searching as a human activity has been widely investigated. There are basically two criteria: the first is the search criteria, that is, the query that the user inputs; and the other is the selection criteria, which is the means of choosing the relevant results (Bental et al. 2000). Dealing with both of them requires cognitive processing.

Regarding search criteria, studies have shown that people do not use many keywords in searching, and Boolean operators, such as AND and OR, are not generally used. Furthermore, many users do not understand the logic behind these operators (Spink et al. 1999). There are numerous challenges for search user interface developers, especially when searching personal content (instead of more or less impersonal web resources) is increasing among the general public. Regarding the selection criteria, the users need to explore the result space, to choose the most promising results, and to evaluate the relevance of the result. Aula (2005) has studied the search strategies of highly experienced, elderly, and less experienced web users. Not surprisingly, the results showed that highly experienced users are more successful in not only finding required information, but also in evaluating the search results. Aula suggests improvements to selection criteria, such as presenting textual result summaries.

An important aspect of human behaviour is *successive searching*. As time progresses, many users search the same (or possibly other) resources for answers to the same or evolving problem at hand (Bateman 1998). Spink et al. (1999; 2001) found that this is also true for web searches. They also noticed that users tend to employ simple search strategies, and that they often conduct more than one search on any one topic, that is, they employ successive searching. Based on their analysis, they suggest that web search engines should provide a way of saving users searches for future use, in order to better support successive searching. Since then, search history has been implemented in, for instance, Google's search toolbar.

The discussion above concerns searching in general, yet in many cases it is targeted at web searches that concern mostly non-personal content. However, searching for personal content is different compared to web searches, especially as the user is familiar with many characteristics of their content, including the context related to it (Cutrell et al. 2006).

To summarize the above brief look at the user's physical and mental capabilities and limitations, it is important to take various individual differences into account in user interface design. The users of the system have different backgrounds and education. They have different levels of experience, some of them being more familiar with the system, while others may have just started using it.

It is also important to take into account language, cultural differences, and requirements for localization, including different meanings of colour and symbols, or another set of alphabets and reading orientation.

6.1.2 User-Centred Design

"Design for all" is a slogan that should be kept in mind when designing a product with broad appeal. Not all of us have fast reflexes or perfect vision, some users having different kinds of disabilities. For example, 8% of males and 1% of female users are colour blind (Dix et al. 2003).

However, the primary prerequisite for successful UI design is to know "Who are the target users, and what the system should do for them."

Although this sounds straightforward, systems are becoming more complex as they evolve from dedicated devices to toolboxes (section 2.2 for discussion on different device categories), as well as having a broader market appeal that translates to even more diverse user groups. To create a usable system for a certain purpose, the designers need to know the tasks a user needs to perform to achieve their goals. Furthermore, for each task the designers should define what kind of steps it requires, in which order the steps should be taken, and what kind of terminology, conventions, routines, external information, and real world artefacts it relates to.

In order to do all this and more, the designers may use a set of design approaches, known as *user centred design* (UCD), which aid in understanding the users and their tasks, and so meet the user's expectations. These approaches are used in different phases of design, some of which are described below. For more discussion on UCD, refer to Carrol (2003); Dix et al. (2003).

Observing and *interviewing* present users with aids in gathering facts and opinions that will assist potential users. Typically, interviews are on an informal one-to-one basis, where the user describes their opinions, routines, concerns, and so on. However, it is extremely useful to observe the users performing the tasks in a real environment (*context*) in order to understand the processes, related artefacts, and information, as well as required phases.

Participatory design takes the actual users along various design phases. Since the designer is rarely the user (or vice versa), it is considered beneficial to include actual users in the design process in order to exploit their domain expertise.

Iterative design and (rapid) *prototyping* refer to creating interactive demonstrations at a fast pace to test different concepts. Based on usability evaluations, the demonstrations may be iterated further or in some cases discarded (validating the design and usability engineering will be introduced in section 6.3). The benefit of this approach is that it enables swift testing of different kinds of design solutions, even if all the other parts of the system are not yet ready. Potentially the prototypes reveal, in early design stages, usability problems that could potentially have a negative impact on the final product. Furthermore, high fidelity prototypes give tangible demonstrations on the concepts to various stakeholders, for instance, management, marketing, and customers, which aid communication between the people involved.

In addition to UI designers and possible users, the design process should involve many other people, such as graphic designers, hardware and software developers, technical writers, and managers, to assure that the design is valid and feasible.

6.2   Interacting with Mobile Personal Content

Besides understanding the general characteristics of people, when designing user interfaces for mobile media, it is equally important to examine the ways they interact with content. As far as digital personal content is concerned, the changes in behavioural patterns significantly affect the interaction. In this section, we will look at interactions that are typical with the most prominent forms of personal content – photos, music, and video.[2]

For changes in interaction, mobility is the key factor: timely information, instant sharing, and spontaneous use patterns are direct consequences of mobile access to personal content. Mobile devices, especially those used for personal communications, are among the most trusted digital devices consumers have ever experienced. This is not because of their technical superiority over other digital devices,

---

[2] The findings presented in this section, relating to user experience and users' behavioural habits with photos, music, and video, are gained in several internal studies over the last six years. So far, the studies have been conveyed in Nokia internal technical reports only, and the results are now publicly available for the first time.

but because of their intimate role. For example, they act as mediators between the users and their loved ones, storing the most important digital personal information.

A mobile device is also a vehicle for *expressing status* and personality. What is interesting is that such issues are not only expressed through features and appearance of the device, but increasingly through the content within it. Those who bring the most fresh, most rare songs to a party are the most respected. Coupled with a desired device, say Apple iPod Nano at the time of writing, this contributes strongly to status among peers. This important aspect of mobile personal content is referred to as *freshness*. "Fresh" is not always equivalent to "new": i.e., an older song may be fresh if it is by an independent artist not publicly known inside the community.

One of the most interesting findings related to both music and photos is that the users do not spend time in file management. As long as the content can be easily accessed, all is considered to be fine. It is only when content access becomes tedious or a desired piece of content cannot be located, that some management needs to be performed.

In essence, the motivation for creating personal content can be categorized into three groups:

1. capturing and storing personal experiences and events;

2. expressing self-identity (flirting, for instance); and

3. enjoying and sharing digital content.

In many cases, the different uses of personal content are not exhaustive, but often interlinked with each other.

6.2.1 Music

As discussed in Chapter 2, mobile music enjoyment has been widely available for several decades. Over time the devices have been miniaturized, but the fundamental experience of music enjoyment has largely remained the same. However, many aspects related to music management have now changed, as music is arguably the first form of popular content that has gone almost exclusively digital.

There are new situations for enjoying music enabled by small digital devices. Mobile players are useful while jogging, skiing, cycling, and even swimming. Some MP3 players weigh just a few grams, making them extremely portable. On the other hand, audio player functions are being incorporated into other devices, such as mobile phones and cameras. We can easily carry such devices into all situations, which allow us to enjoy musical audioscapes as the background to our multi-tasking mobile lives. We can then say that MP3 players are truly becoming ubiquitous.

While digitalization has allowed superior portability, it has also virtualized the actual sound objects. Instead of visible grooves on vinyl, the sound resides somewhere in the abstract bit space. The most profound change has happened in the way we purchase music. Since iTunes opened in 2003, millions of buyers have chosen to buy downloadable music over networks, instead of CDs from music stores. More than a billion songs have been sold so far.

Users of networked music services, such as Pandora,[3] do not even know where the music originates from. While this is only an issue to some die-hard artefact lovers, it nevertheless has rendered our music collections virtual, unreal, and intangible. This means that music management is now prone to the same problems that have plagued file management in other domains, such as corporate document bases, in the past. Owners of large digital music collections need good data management tools to organize their collections – or to even have an idea of what they actually possess.

Besides management, the notion of ownership of digital music is changing. MP3 songs are just like any other files, easily copied and easily sent. We may copy 10 000 songs from our friends with just a couple of mouse clicks. This was not as easy with physical media. We may then consider digital music files as a commodity, not as copyrighted objects. Obviously copy protection changes the situation, but it remains to be seen whether really effective schemes for content protection can be deployed. It is ultimately the post-modern user type we are discussing here, that is, people that are not at all interested in ownership.

People tend to give relaxed interpretations of the ownership of a music object. They may consider as "personal" files that they have downloaded from a file sharing network, or ripped from a borrowed CD. No matter how the actual ownership of music is defined, people still form personal relations with the objects. For instance, my favourite tune playing on a radio channel is "mine" in the sense that I consider it important to me – even if I do not own the object.

Our notion of personal content includes "metacontent", which talks about other pieces of content. For instance, playlists, song ratings, and artist recommendations all serve as qualifications of music objects.

---

[3] http://www.pandora.com/

Even though we do not own the content, we can own the playlists and ratings that we give to songs. Some of the metacontent can originate from commercial or public services, such as the ID3 tags in MP3 files.

Since digital content can be infinitely reused, it encourages sampling – incorporation of pieces of music into new contexts. Currently popular tunes usually incorporate thousands of sound samples, many of them reused from other pieces of music. For instance, a kick drum sample in a typical drum-and-bass song may have been re-sampled by five to ten different artists through as many successive generations of songs.

Other examples of reuse are remixing, which involves taking pieces of music and rearranging and processing them in novel ways. This also includes mashups, which are created by taking two seemingly unrelated songs and joining them into interesting hybrids. Some artists are experimentally releasing their songs in multitrack formats that give users the freedom to remix the songs as they wish. In principle, this transfers some of the production responsibility from the producers to the consumers. While most people may still choose to enjoy music as it is, hundreds of thousands of others are increasingly treating music as digital objects that are just inviting to be modified.

6.2.2 Photos

Prior to the digital photo era, a photo was captured on film by exposing it to light. The captured photo remained inside the camera until processed, in most cases in a commercial photo development lab. This was often done manually, allowing an experienced developer to significantly enhance the technical quality of a photo. Each photo was valuable, since the film could be exposed only once, and often the decision whether to take a photo or not was carefully considered.

Once the photos were developed, they were sorted – failed shots were discarded, whereas the rest were stored, for example, in a shoebox stuck at the back of a cupboard, or in an album with dates and captions. They were then later showed to relatives and visitors for their great pleasure, either by flipping through the album, or by browsing a pile of prints with the number of fingerprints showing an estimate of the viewing count.

And now, photography is getting digital. In 2006, according to one study, some 70% of all photos were taken with a digital camera, while it is estimated that the share will be 90% by 2010.[4] This is a major

---

[4] http://blogs.zdnet.com/ITFacts/?p=10613

change, which is taking place very rapidly. The estimates of digital photos taken during 2004 vary between 20 billion to more than 200 billion. Either way, it is a lot of pixels.

With digital photography, all aspects change. No need to restrict the amount of photos taken, and no need to wait until development to determine the quality of the photo. One could shoot dozens of photos of a single target just to make sure that at least one is successful, and determine immediately after shooting whether a shot is worth retaining or not. Since the cost of storing a single photo is practically zero, there is no need to worry about increasing costs.

There are two significant consequences of this. First, the amount of taken photos increases rapidly. Then the amount of photos taken of a single target increases. This poses new challenges in choosing the best shot out of so many. In fact, user interviews repeatedly reveal that this task is getting increasingly important – and difficult.

Second, the post-processing of photos is more often transferred from a professional developer to a consumer. Our user studies clearly show that most users are not willing to edit or post-process their photos to any extent. The most common task related to editing is scaling the content for different purposes, such as publishing on a blog or sending through e-mail. Beyond scaling and occasional cropping, hardly anything else is done by the majority of consumer users.

Amongst different photographing devices, there are significant differences on the use patterns and expectations of the end result. With a dedicated digital camera, more time is spent on controlling the target and the camera itself. The user often frames the target more accurately, and spends some time on technical aspects, such as choosing a correct program. However, with a mobile phone equipped with a digital camera, the focus is on convenience and social aspects, such as taking a quick shot and then probably immediately sharing the photo with friends (photo sharing is discussed in section 7.3). The communications capability of a camphone is driving the use patterns towards sharing. Furthermore, since a phone is carried most of the time, whereas a digital camera is not, the camphone is used as a reserve recording device when the situation calls for it.

Obviously, photos taken with a digital camera can also be shared. The process is in many cases more complicated – the photos need to be transferred to a PC, scaled down, and then posted on the Web or sent via e-mail.

Interviewing a professional photographer, who recently migrated from the analogue to digital domain, revealed an interesting viewpoint. Instead of carefully framing the target before shooting (as was always done with analogue devices, as long as time permitted), he stated that for non-professional use, current digital cameras contain enough mega-pixels for shooting on the rough and doing the framing and cropping later with photo editing software. It should be kept in mind, however, that this does not concern the majority of users, who do not need to perform such operations on a daily basis.

6.2.3 Video

As a form of personal, especially self-created, content, video is a new addition compared to photos and music. There have been 8 mm film recordings of family events for several decades. However, it was not until mid-1980s, when VHS cameras became available to the general public, that video became a widespread form of personal content.

The way video recording varies from music and photos is the extra post-processing effort that is often required. Even a short video clip of a few minutes is unpleasant to watch in cases where it has not been edited. Shooting a video also requires some experience and understanding of basic principles of filming, in order to produce enjoyable content. For instance, even though zooming in and out is fun, and interesting effects can be generated, watching such a clip is usually unpleasing. In a similar manner, sudden camera movements and improper cuts from one scene to another add to the list of issues to be avoided that a video hobbyist should be aware of. Compared to still photography, more attention and patience are also required.

As with many content formats, video is also going digital. In fact, much of the digitalization in video has already taken place, even though this trend is not evident from a consumer point of view. As a result, the video recording devices are getting smaller, and the imagination would appear to be the most limiting factor as far as editing is concerned. The video can be transferred to a PC, and then edited with a favourite software package. The final result is burned on a DVD that is viewable on any consumer DVD player. Some video cameras store the video stream directly onto a DVD, thus bypassing the need to content transfer, unless some more complicated editing is required.

Once again, mobility has changed, and will continue to change the usage patterns of video content. With the inclusion of cameras capable of shooting video clips, the ubiquitous mobile phones are increasingly used to store not only still pictures but also video feed on daily events and activities. While the storage is at a premium, the resolution is low and the editing options – at least in the device itself – are limited, the popularity of mobile video recording is evident. However, poor quality is an issue that is fading rapidly. A good example of such a device is

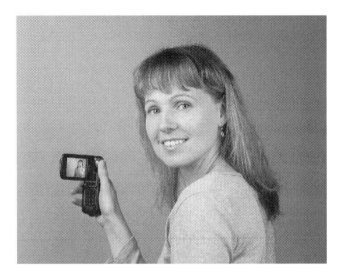

Figure 6-2. Nokia N93 multimedia computer with a high-quality video recorder (Photo: Courtesy of Tero Hakala).

Nokia N93, a multimedia mobile computer with nearly DVD-quality video capturing capability (Figure 6-2).

What is different about mobile phone video capture compared to stand-alone video cameras is the intended use of the clips. The purpose may not be to produce a clip that will be shown to distant cousins or occasional visitors, but rather to store the event for future personal reference, to send out a proof of an extraordinary incident to a friend, or just to experiment – "since the camera happens to be there." Furthermore, mobile video cameras are also used to enhance social presence, and to facilitate group dynamics.

A remarkable current trend related to video content is the increasing interest towards video publishing and sharing. For more discussion on this emerging topic, refer to section 7.3.

An interesting subgenre in mobile videos is mobile micromovies. As the term suggests, this is a very short movie clip. Hannu Nieminen of Nokia characterized micromovies as follows, in a workshop in Tampere Film Festival, 2002:[5]

- a new viewing situation

- free distribution, and

- grass-root level participation.

---

[5] http://www.uta.fi/festnews/fn2002/eng/wed/microeng.html

As discussed in Chapter 2, the *gaps in life* phenomenon is especially suitable for micromovies. The small viewing area and potentially uncomfortable viewing environment do not ruin the experience since the required time is only minutes at most. Grass-root level participation – the ability to contribute to the content creation with built-in video recorders in mobile phones – has further boosted the trend.

6.3   Interfaces for Mobile Media Devices

In the preceding section, we have discussed the interactive system and its user. Furthermore, we have examined some behavioural patterns and changes related to interacting with personal content. Now we will focus on the interface that sits between the system and the user, mediating the communications in both directions. It is the interface that finally determines how the interaction with the actual content takes place, and how well it supports observed behavioural patterns and user habits.

Typically, desktop computers are used in an indoor environment; mobile devices, on the contrary, are frequently used in situations where operating the device is not the primary task being performed. These devices are used either continuously, or in a nomadic manner.

CATHY SCORES A HIT

```
From: catherinedeloitte@get2.net
To: steve.g.mccarthy@example.com
Date: 2008-06-01 19:42:33+0000 (GMT)
Subj: Run over by a truck

Stevie dear, you won't believe this. I was hit
by a truck! Don't worry, I'm okay, did not even need to
see  my  doctor,  but  it  really  could  have  turned  out
worse.

I was just walking down the street in the City, near Crow
Street where we have that gallery, maybe a little carried
away  as  the  old  lady  so  kindly  granted  us  funding  for
another  two  years.  I  was  just  so  happy  that  I  dropped  in
to the Starving Artist and got a ton of new gear, brushes
and  canvas  and  suchlike,  and  then  tried  to  negotiate  my
way back to the Tube.

So Deena called, it's 1st of June so she has fresh cell
phone  minutes,  and  we  kind  of  just  chatted  about  the
```

```
grant, a little difficult with the bags in hand, and then
there was another call coming in which I tried to check
when running the lights at the same time ... you get the
picture. Maybe it's just that I have a one-track mind.

Anyway I dropped a bag and tried to reach it while cross-
ing the zebra, and lost my balance. Hate those high heels.
I sort of staggered about and then the truck appeared
from nowhere, tyres screeching and everything and it
almost managed to stop in time, but did hit me such that
I was just bumped down on the street. No bones broken or
even scratches or anything, but it was scary.

People jumped out of cars and cried and called the police
and everything; everyone was so agitated, I think I was
the only one reasonably calm. The most scared person must
have Deena, as she heard everything over the phone, she
knew something went wrong but did not have a clue, and
she was going all over the place when I finally remembered
my cell phone and found that our call was still on.

So, maybe I'll have to stop running the lights or talking
and walking.

Seems too much for me now. Is it that we are getting old,
Stevie dear ... ?

Cheers,
Cathy
```

As can be seen in the case above, when mobile, the user cannot pay attention to the device's display for long without disastrous effects. This all highlights the importance of user interface and its design.

6.3.1 Why not Speech User Interfaces for Mobiles?

A speech user interface, be it command-based or continuous, could partially solve the problem of continuous use with its hands- and eyes-free operation. In an SUI, the input and output devices do not require as much physical space on the device surface, and they are likely to consume less power than a display with a backlight. The system contains fewer mechanical parts and is easier to manufacture and thus less prone to mechanical failures.

Unfortunately, current technical solutions for speech-only user interfaces are not there yet. For instance, in a mobile system for dictating short messages, native speakers can achieve over 90% word accuracy in moderately noisy environments, such as in a car, after short enrolment session (Karpov et al. 2006). This sounds high enough a recognition rate until one realizes that roughly every tenth word that you speak would go unnoticed, or even worse, would be understood incorrectly in a non-predictable manner. The message would eventually get through, yet a lot of repair and repetition would be required.

What is another aspect is that we cannot expect the users to start training their devices and even if they did, the domain would be limited and language-dependent. For non-constrained speech recognition, there are just too many ambiguities in spoken language that need to be solved and that depend on context and common sense reasoning, for SUIs to take off in the foreseeable future (with speech as the primary modality).

There are also challenges related to mobility. Regarding use context, many environments are hostile to speech interaction. Noisy environments result in recognition errors, and privacy is compromised due to overhearing, these being just two examples of context-related challenges. There are also some social and psychological barriers to overcome before an SUI becomes reality for the general public. For instance, people do not wish to talk to machines. Most likely, though, this is a current trend that will eventually fade, as did "speaking to yourself" with mobile headsets a couple of years ago.

Although the design of speech user interfaces shares common rules and characteristics with the graphical UIs, the whole dialogue between the user and the systems differs from the graphical user interfaces. This is partially due to the fact that audio is by nature a streaming modality (for instance, pausing does not help you to remember the last actions you performed) and hence the speech UI structure is different from GUIs. For example, with a graphical UI the user can see the state of the UI, but on speech UI the audio cannot be frozen. This implies a need for different navigation aids as it is difficult to operate the device when you cannot in any way sense what state the device is in, or what the available actions are. Furthermore, going through a long list of objects is slow, since the user cannot perform a quick visual scan. In addition, the current position in the list is difficult to present.

We conclude that our focus is on graphical user interfaces, yet the advances in speech UI technologies and techniques should be followed closely.

6.3.2 Graphical User Interfaces

To understand graphical UIs for mobile devices, let us look at graphical WIMP user interfaces that have dominated desktop computing for more than two decades. The acronym WIMP highlights the key characteristics of the user interface: windows contain information, menus offer functionality, and the icons present objects graphically to the user. Pointing devices (such as a mouse, a pen, or a touch pad) use the *direct manipulation* interaction style, where the user can point and interact with the visible objects (such as icon or part of the UI) on the screen directly. Windows contain graphical elements known as UI components (widgets), which provide output based on the user's input. The term *desktop* also refers to a metaphor, which attempts to mimic the real-world desktop, where windows represent documents on a table top. A benefit of direct manipulation interaction is that it lets the user be in control (van Dam 1997; Myers 1996; Shneiderman 1998).

In addition to direct manipulation, we distinguish other interaction styles that are highly relevant to mobile devices. The first one is *menu selection* interaction style, which presents all the options available at any particular moment as a list of items and the user can then select the most appropriate one from the list. The benefit of this interaction style is that it provides a clear structure that requires little learning as long as the terminology does not get in the way.[6]

Form filling is an interaction style of choice when lots of data entry is required. Users can typically see all the related fields, navigate between them, and enter the data where required. However, the user must understand the purpose of the fields as well as valid input values. This makes form filling more suitable for intermediate users (Shneiderman 1998).

The graphical WIMP user interfaces of today's PCs should not be used in mobile devices as such. The visual representation and feedback, as well as managing multiple windows, often demands a large display and requires a lot of the user's attention that cannot be provided by mobile devices in mobile use situations. In addition, mobile continuous use is difficult with a pointing device, since targeting at small icons requires accuracy that cannot be achieved easily while walking and using the device with one hand.

6.3.3 Interaction Technologies and Techniques

To interact with the system, the mobile device provides *input devices* for the user to communicate their action and *output devices* that present the system's state and feedback to the user. Often the mobile device is a single entity that contains all input and output devices.

---

[6] It should be noted that menu selection refers to the main style of interaction with the device, whereas WIMP menus refer to user interface components used for providing access to the system's functionality and aid the main interaction method, which is direct manipulation.

Mobility affects the device's physical shape and dimensions, which are together referred to as the *form factor*. Furthermore, industrial design sets requirements and constraints for the form factor as the device should fit on the palm, and should immediately reveal whether the orientation of the device is correct when grasping it. Furthermore, industrial design is affected by ergonomics aiming at maximizing performance, safety, and comfort of use. Ergonomics affect the location and the layout of the physical interface components, such as the display element and the arrow keys, even though the available surface space is often at premium, thus providing few options.

Due to mobile use situations, industrial design should also try to reinforce so-called blind (eyes-free) use by carefully studying the placement, shape, and styling of the input devices. At the same time, the design should prevent unintentional key presses when placing the device in a bag or pocket. As a consequence, various key guard (or key lock) mechanisms (such as hardware switches, button press combinations, and automatic key locks) have been developed.

The most common form factors are monoblock, clamshell (or flip), swivel, and slide (Figure 6-3). Each form factor provides some benefits. In clamshell, the display and keyboard are protected when the device is closed; swivel form factor lets the user view the whole screen even though the keypad is hidden; slide mechanics enable the user to see the screen as well as some input devices; and monoblock does not contain mechanics that could be prone to mechanical failures in rough handling.

Compared to desktop computing, where the display, keyboard, and mouse dominates the interaction hardware, the mobile devices are more heterogeneous. Before discussing how to use the various hardware components for interaction, let us review them quickly.

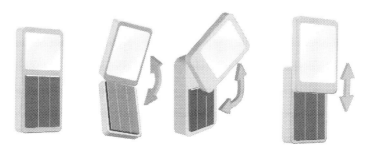

Figure 6-3. Illustrations of mobile device form factors. From left to right: Monoblock, flip (clamshell), swivel, and slide.

6.3.3.1 Output

Depending on various aspects, such as device capabilities, user's task, and the current environment, there are many ways to use one or more output modalities, even though most mobile devices rely heavily on presenting information in visual form. Also, the main presentation modality can be enhanced with the support of other modalities (Dix et al. 2003).

Considering the mobile user interfaces, a principal device for presenting information is a display. Various technologies ranging from simple black-and-white LCD to high-quality colour TFT displays exist. During recent years, displays have improved significantly in terms of colour-depth, resolution, and quality, but still their size is a fraction of a typical desktop computers' display. This is a factor that has a significant impact on the graphical user interface design. Figure 6-4 visualizes this by depicting relative sizes of the following display resolutions:

- 128 × 160 pixels (Nokia 6111 and SonyEricsson Z600 mobile phones);

- 220 × 176 pixels (Motorola Razr mobile phone and Magellan eXplorist XL GPS receiver);

- 320 × 240 pixels – Quarter VGA or QVGA (Nokia N92, Motorola A780, SonyEricsson P990i smart phones; Archos AV400 portable video recorder);

- 640 × 480 pixels – a standard VGA resolution;

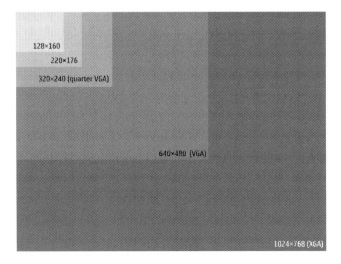

Figure 6-4. Relative screen sizes of display resolutions used in various mobile devices.

- 1024 × 768 pixels (XGA) is a typical resolution used in current mainstream laptop computers.

There are also more primitive ways of conveying visual information. LEDs, for instance, are used for providing simple feedback related to the device state, such as battery level that can be indicated with different colours or frequency of flashing.

Audio output can be divided roughly into two categories: speech and non-speech. Speech output is based on text-to-speech synthesis, or playing files that contain speech. Text-to-speech (TTS) synthesis is used, among others, in the Nokia N91 multimedia computer for playing recognized voice commands, or speaking aloud the caller's name.

Several devices are capable of producing mono- or polyphonic tones and playback audio files, which make the use of audio versatile and appealing. Notifying the user about an event or providing feedback on operations, such as key presses, are typical examples of using non-speech audio output.

A prominent difference between mobile and desktop UIs is the use of vibration as an output modality. In desktop computers, vibration (for example, via a vibrating mouse) has not gained popularity.[7] However, phones using vibration for notifying, for instance, about an incoming call have more appeal. The benefit of tactile feedback is that in a noisy and busy environment the user does not have to pay attention to the device constantly, but may place it in pocket and only take a closer look on sensing vibration.

A central concept related to output is feedback that refers to the system's response to the user's action. It can be based on one or more modalities. For example, when the user is typing text by pressing the keypad, the key press is indicated with a small audio beep while the typed characters and the cursor position are displayed on the screen. In addition, each action that the user performs should give feedback in some way. For frequent actions, the feedback should be subtle so as not to disturb the task flow. Feedback is also something that users will appreciate when interrupted by something sudden and unexpected.

6.3.3.2 Input

As the mobile UIs are descendants of graphical WIMP user interfaces, the input devices related to pointing, selecting, and entering text are the most relevant.

---

[7] Currently the only domain in desktop computing where the vibration is used is in gaming with the devices, such as racing wheels and joystick with force feedback.

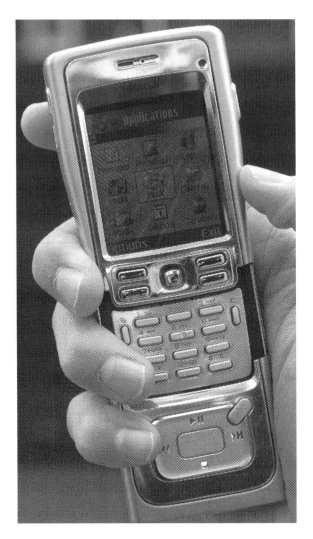

Figure 6-5. Nokia N91 with two softkeys labelled "Options" and "Exit", ITU keypad, 4-way joystick with centre select and dedicated music keys (Photo: Courtesy of Tero Hakala).

As the number of physical keys on the surface of the device is limited, it uses softkeys to bypass this limitation. A *softkey* is a hardware key located near the screen, with its label visible on the edge of the display. The function of the key, such as opening a menu, and its label change depend on the current state of the device. Figure 6-5 shows two softkeys below the display. In addition to mobile devices, softkeys are ubiquitously found in ATMs around the world.

Figure 6-5 also shows an example of a keypad for text input. The keypad is based on a standard and consists of numbers 0–9 and characters # and * (ITU). To input text, three or four characters are assigned to each number key. For example, to type the letter "B" the user presses the key "2" twice, since letters A, B, and C are assigned to this key. Other text entry methods include those based on predicting the word that the user is typing, such as T9.[8] Also, other keypad layouts, QWERTY among others, exist.

If a device is not equipped with a full QWERTY keyboard, the mode of the character entry should be displayed. For example, does a key press in an ITU-T keyboard generate a number, a character, or a symbol? To facilitate eyes-free use of the keys, they often have markers or distinctive shapes. For example, in ITU-T alpha, the 5 key has a physical "bump" that is recognized by the finger.

In addition to the numeric keypad, Figure 6-5 also shows an example of dedicated media keys on the cover of the slide that are used for controlling music playback.

Most mobile devices must have an input device for moving a selection or pointing to an object on a screen. For discrete input, they usually employ a 2- or 4-way joystick, a rocker key,[9] or cursor keys. Half-cardinal points, such as north-east, are not typically supported since they could produce unintentional navigation steps when the device is operated while moving.

A 4-way input device may contain a so-called *centre select*. Once the centre of the device is pressed down, the device effectively transfers into a 5-way joystick (Figure 6-5). If they do not contain such a device, a soft key or a dedicated hardware key is used for selection.

Another solution for changing the selection, in a way that the user can adjust the speed, is by using a scroll wheel; the selection can be confirmed by pressing the wheel in. The wheel is used by forefinger or thumb, depending on its placement that varies from the front (as in Apple iPod) or to the side (as in RIM Blackberry or SonyEricsson P990i). In Apple iPod, on the contrary, Click Wheel is a dual purpose input device. If the user rotates a thumb on it, the wheel can be used for scrolling, but if the user taps cardinal points near the wheel's edge, those locations act as buttons (iPod; RIM; P990i).

For continuous input, some mobile devices are equipped with a touch-sensitive screen to move a cursor and select objects with a stylus (a pen) or a finger. To input text, the user can draw the characters directly on the screen. In handwriting recognition, the system interprets the pen gestures as characters and sends them to the application as

---

[8] http://www.t9.com/
[9] Rocker key combines several cursor keys to a single key that can be tilted (rocked) to left, right, down, or up when pressed in that direction.

text. In some cases, such as Palm's Graffiti (Palm b), the drawing of characters is simplified and they can be formed with one stroke of the stylus.

Devices with a touch sensitive screen also support so-called virtual keyboards, where a keypad is drawn on the screen and the user can tap its keys with the stylus or the finger (7710; Palm; P990i). In addition, handwriting recognition can be used for distinguishing gestures drawn on the screen in order to perform a predetermined action. Also the device's embedded camera may be used for recognizing handwritten or printed characters (Motorola Ming).

Recently, using the mechanics of the device as input has become commonplace. For example, as the user opens the clamshell device and twists the screen, the device will go from the idle state to the capture image state, as with Nokia N93 (Figure 6-2).

Different kinds of sensors may also be used for input. For instance, devices targeted at sports often contain embedded motion sensors, such as a 3D accelerometer for monitoring the step count. In many cases, the accelerometer also doubles as an input device for detecting some predefined gestures, such as tapping the screen with a finger. Similarly, GPS satellite positioning provides implicit input about the user's current location (for more discussion on GPS and location, refer to section 7.2).

Selecting, or pointing or clicking an item is often a prerequisite for further interaction. There are two primary pointing methods: direct and indirect. In direct pointing, the user can point an arbitrary single spot on the screen, for instance, with a stylus and a touch screen. In indirect pointing, the user manipulates a separate input device, such as cursor keys or a joystick, in order to move the visual selection indicator. As a consequence, indirect pointing requires good hand-eye coordination and cognitive processing, since the brain has to interpret the changes on screen and control the hand at the same time (Shneiderman 1998).

Selection can be done by hardware keys, such as a softkey, a dedicated selection key, centre selection in 5-way joystick, or a stylus. There are also ways to enhance the selection and associate different operations with it. For example, the operation linked to a button differs depending on how, or how long, the user presses the key. The most common selection methods are a single press, a long press (the key is pressed down for more than a predetermined period of time), and double press, where two consecutive single presses are made during a certain period of time.

Occasionally, the user needs to select more than one item, referred to as multi-selection. With an indirect pointing device, this task requires the user to switch to a special selection mode, where they can toggle several selections on and off independently. Mode switching is performed via a menu item, or the user may press down another key to enter in multi-select mode.

6.3.4 UI Structure and Navigation

Graphical user interfaces for desktop computers, such as Apple MacOS or Microsoft Windows, provide *application windows*. These windows divide the screen into regions containing application-specific information and controls; the user can also switch between the open windows to change the active task quickly and easily. Displaying several overlapping and resizable windows at the same time requires a large display in terms of physical display area and pixel resolution.

Since mobile devices have small screens, presenting multiple windows in a similar manner is problematic. First, the controls needed for window manipulation, such as borders and a title bar, consume screen real estate. Second, as mobile devices rarely have continuous pointing devices, direct window manipulation and moving between windows is difficult and slow. Therefore, instead of using the term application window, Symbian OS-based Series 60 UI refers to an *application view* as a collection of user interface components that the application requires.

The screen layout is divided into regions for dedicated purposes. The topmost region, the *status pane*, presents the status of an application and/or the device; the main pane, at the centre, contains the application data; and the navigation pane at the bottom displays softkey labels, as shown in Figure 6-5 (Symbian 2005).

Since available screen space is at a premium, mobile devices cannot present information in a parallel manner. For instance, an MP3 player first shows all available artists in one view. After selecting an artist, the user moves to another view that presents all albums by the selected artist. Then, once the desired album is selected, the user is able to browse the tracks, once again in another view.

The example above illustrates sequential information presentation, where the views are separated and the user must make a selection before moving to the next view.

A sequential displaying of views forms a hierarchical tree, where the user moves (navigates) between nodes that present the views, sometimes called "states". Figure 6-6 shows an example of an overall navigation structure of a hypothetical MP3 player. The first state becomes active after the device's power has been turned on, and thus it is often called the idle or main state. From the idle state, the user can enter into one of the three states that are descendants of the idle state (Music library, MP3 player, or Settings); each of them has further sub-states.

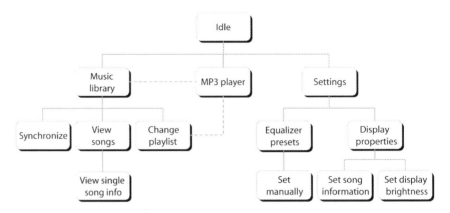

Figure 6-6. An example of navigation structure.

The dotted lines in the figure represent additional navigation shortcuts between the nodes to speed up frequently performed operations. The shortcuts effectively transform the UI structure from a tree to a graph.

Figure 6-6 demonstrates that even a simple media device can contain several views and each of them can have further sub-views. This brings up an important topic: how to actually navigate between views?

Put briefly, in navigating between the views we can use menu selection interaction style[10] that was popular in desktop computing before the WIMP GUI era. In menu selection interaction style, the user selects the most appropriate of the available options shown on the screen. The selection will then cause a transition into a new state (or view).

There are numerous ways of performing the selection and transition. Depending on the device in question, the transition is done by selecting an object or an option on the screen, by turning a knob, sliding a switch, or pressing a button. For example, pressing a button to power up an MP3 player boots the device and changes its state to idle.[11] Pressing a play button takes the device to a new state (play), where the playback begins.

Obviously, in addition to selection, the user needs to have an input device that allows moving between the nodes on the same hierarchy

---

[10] Please notice that menu selection here denotes the main style of interaction with the device, whereas in WIMP GUI menu refers to user interface component used for providing access to the system's functionality.
[11] Depending on design, the first state after boot may also be "play" in an MP3 player.

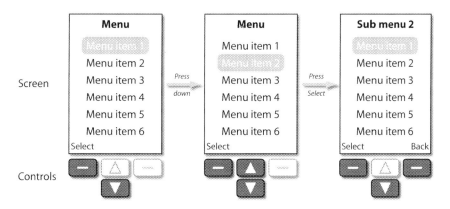

Figure 6-7. Main menu hierarchy (left), moving down (middle), and a new menu list after selecting the second item (right).

level, as well as a way of returning to the previous state, if applicable.

Menu selection is a popular interaction style in mobile devices, such as mobile phones, MP3 players, and GPS navigators. First, it requires little learning, provided that moving from one view to another is clear and the user is familiar with the terminology. Second, the use of the menu hierarchy allows adding new views and features easily. And third, using a menu requires only four keys (Figure 6-7): two keys for changing the highlighted item (navigating back and forth in the hierarchy), one key for selecting an item, and one key for cancelling the selection. This makes menu selection particularly appealing to mobile devices.

In Figure 6-7, the leftmost picture displays a case when the first item is highlighted on the main level, thus the user cannot move up or go back (these buttons are not active). As the user navigates down, the control for moving up becomes available (the middle picture). Should the user select the second menu item, the next hierarchy level will be displayed (the rightmost picture). The title of the view displays the current hierarchy level.

Obviously, menu selection interaction style is not without drawbacks, such as the one illustrated in Figure 6-7. Due to the small screen, only one level of the hierarchy is visible at any one time. Moving to another view reveals the new sub-menu, which hides the previous level, and thus increases the user's cognitive load as they have to remember the preceding selections, understand how the visible items relate to them, and anticipate what kind of items are revealed from the next view(s) (Shneiderman 1998).

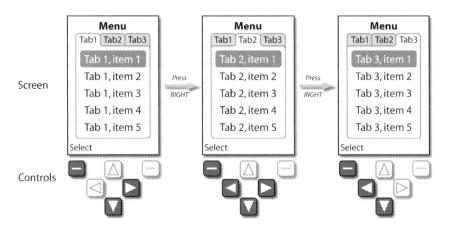

Figure 6-8. Using tabs for moving between sibling views.

Sometimes it is possible to combine related views into a single state. *Tabs* are a means of presenting multiple views and navigating between them inside the same state. Hence, it is the question of moving from one node to its sibling within the same branch. Tabs are useful for presenting multiple content types, where each type can be shown on its own tab. Since each view is displayed as a list, horizontal movement with a 5-way input device is used for changing the tab, as shown in Figure 6-8.

At the beginning of this chapter, we claimed that mobile devices are becoming harder to use, due to the feature race. To be precise, adding features alone does not make usage more difficult, but rather the hierarchies needed for menu selection interaction techniques have become too large and complex. They break the design guidelines for menus, developed in the 1980s (Norman 1991) and consequently contribute to the disintegration of the user interface.

To facilitate learning, the UI structure should be static. Therefore, if some states or sub-states are temporarily not available in the menus, they should not be removed but rather disabled (Figures 6-7 and 6-8). Altering the structure would hinder the navigation since the user may not understand why the location of some states is changed, and why some options are not visible.

In addition to application views, the system frequently needs to alert, notify, present further options, or query input from the user. In order not to disturb the task flow by changing the view, the system makes use of pop-up dialogues, displayed on top of the application view (Figure 6-9). Since mobile user interfaces avoid window management,

Figure 6-9. Examples of modal pop-up dialogues for confirmation and text entry.

the dialogues are modal, as oppose to modeless, which implies that they must be closed prior to returning to the previous view.

Dialogues imply that mobile user interfaces are not based on a "pure" menu selection interaction style, but also contain some characteristics of WIMP. Allowing multi-tasking by the means of switching between application views, and providing the "Options" menu to access functionality, are other examples of this (Series60.com 2005; UIQ 2002).

Mobile platforms, along with their style guides, aim to create navigation mechanisms that are consistent and standard across all applications. Nevertheless, navigation remains one of the most challenging aspects in a mobile user interface. This is mainly due to operational inconsistency: in some cases interacting with a UI component results in a navigation step, but in some cases the same component causes an action. What is most important is to indicate whereabouts the user currently is in the navigational structure, and what are the possible steps the user can take.

Some methods for making the navigation easier include labelling the state, disabling the functions that are unavailable at that moment, and displaying visual cues that help recognize what choices will cause a navigation step. For instance, in desktop GUIs, a menu item that will open a dialogue window is denoted with a menu label followed by an ellipsis, such as "Open . . .". Another way is to label the states and show the related sequence of labels; this method is somewhat analogous to showing the folder structure in a file manager. This technique is also known as a *bread crumb trail*, popular in web user interfaces.

The system may provide shortcuts, accelerator keys, to speed up navigation. For example, in a Series 60-equipped smart phone, a three-by-four grid is used to display the installed applications. Each grid contains an icon that either launches an application or opens a sub-folder, in a three-by-four grid format. In addition to selecting the desired icon with the cursor keys and a selection key, each grid item is also mapped to a respective key in the ITU-T alpha keypad (for instance, pressing the key labelled "4" opens the first item on the second row). This accelerates navigation, since the user does not have to first highlight the desired item and then select it.

ZoneZoom application takes the idea of using the keypad further by generalizing the method for 2D view navigation (Robbins et al. 2004). Each view is divided into nine segments that can be accessed with the keys from "1" to "9", respectively. As the user selects and zooms in on the segment, they can re-segment it with the hash key ("#") or zoom back to the previous view with the star key ("*").

If navigation in the hierarchical structure is problematic, how can it be supported? Most mobile computing platforms provide navigation aids, such as highlighting the item that has the input focus, providing a scroll bar for indicating current position in the list, labelling the view, and displaying the index number of the current menu item. For instance, 4-3-2 denotes that the user has selected the fourth item on the main level, the third from the next level, and now the focus is on the second item of the current view.

Another option is to divide navigation into two: navigation between states, and navigation inside a state. For instance, Space Manager mobile document management system provides different views to navigating between folders (navigation level) and inside a folder (folder level) (Hakala et al. 2005a). Space Manager attempts to maximize the amount of visible folder structure while avoiding the need to pan and scroll. The navigation level (Figure 6-10) shows the folder structure as a tree on a tilted plane. Tilting adds to the sense of depth, thus making use of strong human spatial memory.

At the navigation level, the highlight can be moved to any neigh-bouring folder by using the five-way joystick. Moving up highlights the closest folder on the next level and moving down returns to the pre-vious level. Moving left or right navigates between folders on the same level. If all folders do not fit on the screen at once, the view is automatically panned as the user moves near to any edge of the screen. Selecting the highlighted folder opens it and initiates animated transition that emphasizes the navigation step to the folder level (Figure 6-11).

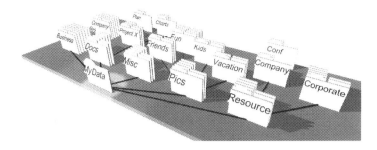

Figure 6-10. The navigation level in Space Manager.

Figure 6-11. The folder level in Space Manager.

When entering the level, the folder opens like a box with the lid open and facing the user (Figure 6-11). By clicking up, the user starts browsing the files in the folder, while clicking down closes the folder and zooms out to the navigation level. Moving left allows browsing the folder contents as a list, and the right button activates the search tool.

What if we do not wish to separate browsing between views and inside a view? This would require us to display all the relevant content objects in a single view, which would require much scrolling. To prevent this, we need to provide a way to constrain the number of content objects in a view. Media Tray photo browsing application enables the user to specify the visible content based on time and content type (Hakala et al. 2005b). Navigation is then divided roughly between the content visualized on the tray, and controls to manipulate the view (Figure 6-12).

Figure 6-12 shows possible navigation paths as pipes and the user's selection is moved as "a ball" in them. The controls for manipulating the view are presented by ball joints in the pipe. This way the user always knows where the focus is, recognizing the possible navigation paths and interactive components. To browse the content, the user navigates from the pipe to the tray. The item closest to the ball joint from which the jump was made is framed with red. Similarly, as the

Figure 6-12. The Media Tray application.

user exits the tray, they move to the edge of the tray and jump to the pipe, even though there are no joints visible.

What if a single object is too large to be displayed all at once? For instance, the MiniMap Internet browser illustrates a way to aid in-document navigation. The normal view shows the web document in its intended layout, meaning that the document is only partially visible. As the user starts to scroll the page, the view is automatically zoomed out to see an overview to the whole document and the current location in relation to the whole document (MiniMap).

6.3.5 Basic UI Components for Mobile Media

We have provided overviews to user and interaction technologies as well as discussed how to structure UI views or states. Next, we present the way some commonly used UI components look and behave from a personal content point of view. We concentrate on visual techniques in the form of text, picture(s), or their combination.

Visual UI components are interactive or non-interactive. Non-interactive UI components, developed purely for presentation purposes, inform the user about the states of the application and objects and about ongoing operations, yet they cannot receive any user input.

6.3.5.1 Non-interactive Visual UI Components

Figure 6-13 depicts basic ways of presenting a content object visually: a text string, an icon, an image, or a combination thereof. Occasionally information may be displayed as an abstract graphical symbol, such as a triangle or a circle. The difference between a symbol and an icon is that the icon presents its object by resemblance, whereas the symbol denotes its object by a convention or a common agreement; for example, the arrow symbol (▶) is generally used for presenting a playback action, and stars (★) are be used for indicating rating (Mullet and Sano 1995).

It is important to consider the size of the visual objects. Small objects do not consume excessive amounts of screen space, but they should still offer sufficient details and distinctive features, so that the user can differentiate between them. Furthermore, the size depends on available input techniques: the size of the thumb affects the size of an object on a touch screen. Size, in addition to other visual attributes, such as colour, brightness, shape, texture, blinking, and orientation, may also be used for encoding additional information into the object.

Text can be presented in one or more lines, depending on available screen space. When screen space is limited, the text string can be truncated, which is often denoted by an ellipsis. The other option is to animate the text so that it scrolls slowly according to reading orientation, so-called ticker tape.

Combining an image and a text string provides additional information, also enhancing recognition and learning. For example, a thumbnail-sized image may be accompanied by the date of creation as a text string, or a menu command displayed as an icon and a text label.

Time can be visualized as a text string or as an analogue clock with face and hands. With textual presentation, exact time is easy to present, but the way it is presented varies from country to country. For example, both the string "18h00" and the string "6.00 pm" denote the same time of the day. Similarly, a date can be presented and interpreted in a number of different ways. The designer should be aware of different calendar systems and notations (Figure 6-14).

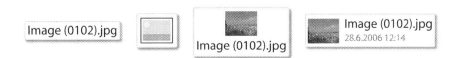

Figure 6-13. Graphical presentation of a single object.

10/07/06   07.10.06

Figure 6-14. What day is it? Different ways to display the date "October 7th".

Related to personal content, summarizing multiple objects and presenting them as a single object are needed to denote a collection or group. For instance, whereas a folder symbol presents a collection of files, multiple albums of an artist could be presented with a single object. There are many choices available: for instance, using a number in parenthesis after the artist name to denote the number of albums, or presenting a stack of albums as an icon.

Dynamic information is presented as video or animation but they should be used with discretion. The playback sequence should be short enough not to disturb the task flow, while still conveying the important information. Furthermore, the sequence should show the start and end points clearly, even if it is looped.

One purpose of presentation components is to notify the user of changes in the system or application state. A concept close to notification is indication, which relates to presenting the current state, mode, or status of the system. Notifications and indications may also be dynamic: for example, a battery level indicator in a mobile phone will change its appearance according to the current power level.

To fit all required information and controls on the screen, the designer has ways to preserve the limited screen space. They may use different visualization techniques, such as perspective distortion, to make room for additional objects (Lehikoinen and Aaltonen 2003). In Figure 6-15, slightly distorting the image and image list has freed some screen real estate for a menu.

Another way to distort the view is known as the fisheye lens, which enables seeing an important object in detail, while preserving the broader context in which the object belongs. An example of this technique is the rubber sheet view that is suitable for small displays (Sarkar et al. 1993).

Figure 6-15 introduces a visualization technique that uses more than two dimensions and thus provides an illusion of depth, which exploits the user's spatial memory. In the figure, a 2D tilted plane is floating in a 3D space. As a consequence, the interaction remains two-dimensional and can be performed with, for instance, a regular five-way joystick (Hakala et al. 2005a; Drucker et al. 2004). Using true

Figure 6-15. A menu, an image list, and a preview of selected image are fitted into a single view with the aid of perspective distortion. This is a synthesized image.

3D visualizations for mobile devices, where navigation takes place along all primary axes (x, y, z) are still rare, partially due to limitations of the graphics performance and lack of six degrees-of-freedom input devices.[12]

Not all screen space preservation techniques require transforming the view. For example, Kamba et al. (1996) introduced user interface components that are hidden until they are needed. When the UI component becomes visible, it is partially transparent so as not to hide any relevant information.

In graphical user interfaces, the currently selected object needs to be identified. A common practice is to highlight the object in question by the means of, for instance, a bounding box. In order not to consume pixels for indicating the selection, Space Manager (Figure 6-16) introduces a flashlight metaphor. The metaphor is based on a light beam targeted at the object of interest. Instead of altering the background colour or drawing a frame round an object, only the brightness of the pixels is varied (Hakala et al. 2005a).

It is not necessary to try and squeeze all the relevant content and information on the screen at once; the user may pan the view instead. When they notice interesting information, they may zoom in to get further information. Essentially, this technique refers to zoomable user interfaces (ZUI), which have been studied for desktop computing (Perlin and Fox 1993; Bederson and Hollan 1995).

Rapid serial visual processing (RSVP) is a browsing technique for a set of images, where one or more images are presented at the same

---

[12] Visual 3D interaction is also rarely seen in desktop computing, games excluding.

Figure 6-16. A flashlight metaphor used for highlighting the "Doc" folder.

time, for a short period of time, after which the next set is displayed automatically.

6.3.5.2 Interactive Visual UI Components

As can be concluded, there are many ways to present information and provide feedback, and we need to know how to interact with the information. In addition to input devices, we also need their virtual counterparts, UI components known as *controls*. Usually, some of the most frequently performed actions are directly linked with a dedicated hardware button, such as those for adjusting volume in an audio player. However, some actions are performed by manipulating the controls on the screen; therefore, the audio player could also have push buttons on the screen for the same purpose.

The purpose of controls is to provide the user with a means to manipulate objects in the application and thus perform different operations. If the view consists of more than one control, one of them has an input focus, implying that it will receive all interactions performed on an input device. Controls have different states, such as normal, highlighted (input focus), selected (pressed), and disabled.

Figure 6-17 presents an imaginary user interface of a minimalist audio player, operated with a five-way joystick. The UI consists of a label for information about the artist and the song, visual push buttons for playback controls, and a slider for changing the playback position within the current track. By moving up and down, the user can choose to manipulate either the playback controls or the slider. Moving left and right adjusts the current track position, or highlights one of the playback control buttons, depending on whether the slider or the buttons have the input focus, respectively.

A push button is one of the most familiar controls in graphical user interfaces. It is designed for performing an action indicated by a textual or graphical label. In some cases, the button also provides access to a

Figure 6-17. A user interface of a simple audio player.

Figure 6-18. A schematic UI for making a search query.

toolbar or menu, which is opened when the button is pressed. A slider, which is similar to a scroll bar used for scrolling the view, accepts a value within a designed range as an input. Since the minimum and maximum values are known, the slider is able to visualize the current value in relation to the whole range. The problem with slider input is defining the proper increment (step size). If the range contains too many steps, large changes become tedious with an indirect pointing device, while too few steps do not allow fine tuning of the value.

Figure 6-18 shows a small illustration of a UI for making a search query, with several types of controls. It has a dropdown text box, where the user can either type the desired keywords, or open the dropdown list for quickly accessing search history (for more information on lists, refer to the next section). The content types the user wishes to search for are presented as checkboxes. However, using this control poses a problem – what if the user does not select any content type?

The first solution is assuming that the user does not care about the content type, and searches for any type of content. This is an illogical choice, though, since it implies that "none equals any". Another solution would be to display an error message that indicates that no content type is selected. Nevertheless, the best solution would be to first use two radio buttons (see below) for setting the content type to be "Any" or "Specify". If "Specify" is selected, the content type checkboxes are enabled. The form also contains a slider for setting the minimum or maximum size of the content object with the aid of a radio button group.

For alphanumerical (that is, text or numbers) input, a *text box* control is used for entering one or more lines of text. Some controls allow the user to style and format the text. In addition, some text boxes provide a feature for entering a password, in which case the entered characters need to be masked by using a neutral character (such as "*") that will not reveal the content. Furthermore, if the text box has a certain range for values or specific format, this should be presented to the user in order to prevent errors and minimize memory load. An ideal way to indicate the format is to provide some default value that the user can then change. If necessary, the user input should be validated and in case of an error, appropriate feedback should be given.

Radio buttons are used for choosing one selection from among many options. Should the user be able to choose an option independently of others, checkboxes are used.

Text input is required in many tasks. Figure 6-19 illustrates a special case: controls for date entry. A date is entered by using a formatted text box (top row), text boxes with spin buttons (middle row), and a dedicated date control (bottom row) that shows a monthly view when opened.

Spin-buttons are often attached to text boxes expecting numerical input. They allow increasing or decreasing the value of the control by predefined steps. This is useful when typing a new value would require more effort than clicking a spin button, or when the user can go through a fixed set of predefined values. In essence, a dedicated date control can be visualized as a push button right next to the text box. Clicking the button pops up the date control. Typically, the control presents time as a monthly view (a grid), where the user may change the month and navigate across days.

Input controls may be grouped together into a single view, known as a *form*. Not surprisingly, forms are related to the *form filling* interaction

Figure 6-19. Three examples of UI controls for entering a date.

style, which is especially useful when large amounts of data input is involved. Figure 6-18 is an example of a form for searching content.

Another way to combine and manage several controls is to use a *toolbar*. A toolbar is a container that can include various controls to emphasize frequently used commands and provide fast access to them. The location of the toolbar may be adjustable, but within given limits. A toolbar that floats on top of the view and can be freely adjusted is referred to as a *palette*. However, the use of indirect pointing devices make palettes undesirable, since they share similar drawbacks to resizable and movable windows in general (section 6.3).

A menu is a convenient way to present available actions. If the platform provides several menus at the same time, they are usually presented in a menu bar that shows the menu titles. Selecting a menu title opens the menu, which then presents the items as a list. Each menu item presents one command that the user can select in order to perform the action. In some cases, a menu item contains a sub-menu that is typically indicated with a "▶" symbol following the textual label.

6.3.5.3 Grids and Lists

So far, we have presented controls that enable the user to input information or perform an operation. Sometimes we need to have a UI component that manages both presentation and control behaviour. The most common components are lists and grids, used to present multiple objects and provide a method to navigate between them.

Perhaps the easiest way to present a set of any kind of textual or graphical objects is by a *list* (Figures 6-20a and 6-20b), where the

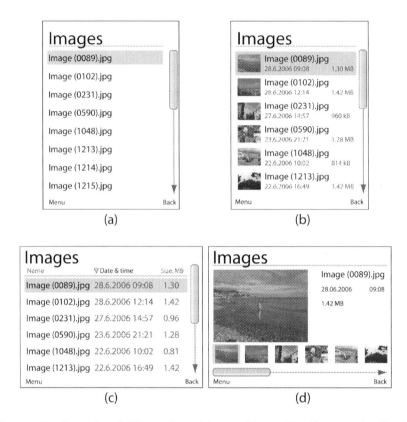

Figure 6-20. Examples of different lists of image objects: (a) with text only, (b) with thumbnail images, (c) with multiple columns, or (d) horizontal orientation.

user can navigate back and forth between the items. A single list item may consist of graphics and multiple rows of text; a list may also have more than one column (Figure 6-20c). The column titles may provide fast access to alternative ways to sort the list. The orientation of the list is typically vertical instead of horizontal (Figure 6-20d) and its layout may vary. If all the list items do not fit into the view at once, a scroll bar is used to indicate the location of the currently visible range of objects.

All lists in Figure 6-20 use the same resolution and font size. What is the best presentation depends on the user's task. For example, the plain list (top-left) presents the largest number of items; the list at the top-right shows more details and a small preview; bottom-left offers a quick way to scan details and change the sort order; and the list in bottom-right shows a large preview of the currently highlighted item.

A special case for presenting a list is a *dropdown list* that requires little screen space, as only the currently selected item is presented. If

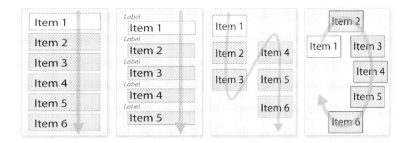

Figure 6-21. A list can have a straight, curved or bent axis.

the user wishes to change the selection, they open the list by clicking the arrow button attached next to the list item. After locating the desired item and selecting it, the list is closed and the chosen item presented. Not only lists can be dropped down: in Figure 6-19, the calendar's monthly view can also be dropped down.

There are also multi-selection list boxes that allow selection of more than one item. The selection is usually done by toggling on and off a checkbox attached to each list item. Alternatively, a mode switch, such as a menu command or a hardware key, can be used. Selected items are then identified with a mark, often a "✔" character, after their label.

A list is a one-dimensional structure; in other words, it has one axis. The axis does not necessarily have to be a straight line, as it can be curved or bent (Figure 6-21). Still, the navigation is preserved in one dimension.

Often, only a small fraction of the content is visible at any one time on the small screen. As a result, the user may not know what choices are available without scrolling the view, which is a tedious task if the list contains a large number of items. Obviously there are techniques for aiding interaction with the lists. For example, the user may sort or filter the list based on some keyword or other criteria.

If another visual axis is added, the presentation is called a *grid*, or sometimes a matrix or a table. The graphical content, such as images, is often shown in the form of a grid. In addition to graphics, a grid item may also contain one or more rows of labels. Since the visual structure and navigation space are two-dimensional, the user is able to see more content and access it with fewer navigation steps as when compared to a list (Figure 6-22).

If the content objects form a hierarchy, it is typically presented with a *tree view* (Figure 6-23). A hierarchy related to music could consist of, for example, artists, their albums, and tracks in each album.

Figure 6-22. An example of an image grid without labels.

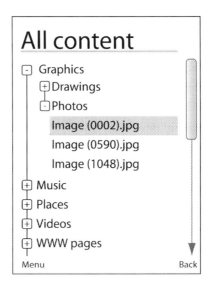

Figure 6-23. An example of a tree view.

Presenting content objects that have temporal relations is straightforward by using conventional calendar views. In Figure 6-24, for instance, a day is presented as a list or a grid for displaying a week, that are familiar and the interaction is suitable for mobile devices.

The rightmost view in Figure 6-24 illustrates a problem related to the use of empty screen space (no events in the calendar). If the days

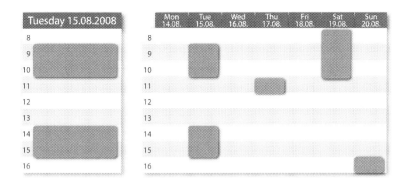

Figure 6-24. Calendar views for a day (left) and a week (right) showing events.

without events are hidden from the timeline, the screen space is used more efficiently, but still the screen may contain several empty areas. On the other hand, we cannot combine several non-related events of a day together, because something important may be hidden. Furthermore, in some cases it is desirable to see the empty slots in the calendar, such as to plan future activities.

An example of a condensed calendar view is the Lifeblog application. Navigation is based on a timeline, where days that do not contain any objects are discarded. For other days, the application groups objects of the day together and presents all of them in temporal order in a three-by-two grid. If the day contains more than six objects, moving to the left or right shows the next set of objects before moving on to the next day (Lifeblog).

Context Comic (Aaltonen et al. 2005) has a different approach for presenting content in temporal order. Instead of showing all single events, Context Comic groups and summarizes related items together and produces a comic-like overview of the most important events during a certain period. Context Comic uses a folded timeline (instead of linear), which moves from left-to-right, top-to-bottom along the entire screen as in a comic strip. Each panel of the comic is a summary of several objects and the size of a panel reflects the amount and importance of the information.

In a comic-like presentation, the panels are in chronological order without uniform interval, which makes the use of screen space more efficient. Furthermore, the view aims at containing only relevant information and if the user wishes to see more information related to a panel, they select the desired panel. Selection opens a new list view that displays all related objects.

6.4   Designing a Mobile User Interface

In the previous sections, we have reviewed and discussed users, technologies, and techniques for mobile user interfaces. The dialogue between the user and the system takes place only through the user interface; therefore, UI design has a crucial role in rendering the whole system easy and efficient to use. In this section, we will look at how to select the right technologies and how to use them, resulting in a first-class mobile user interface.

From the UI design perspective, we are interested in defining what the system should do, and the related environment. We will also describe how the system provides the desired functionality by capturing functional requirements and identifying the required technologies, which are then used to further refine components during the detailed design phase.

```
COMMENT: FUTURE MOBILE DEVICE AND UI DESIGN APPROACHES

Many trends point to the changing nature of mobile devices.
Significant computational and data storage enhancements
matched with the ongoing human need for greater mobility
within our daily activity will provide new opportunities
for mobile device and user interface solutions. However
to ensure that these devices connect with our daily needs
and become integrated into work and leisure activities,
new approaches to the design of these devices and their
UIs will be required. Therefore it is suggested that the
domination of a pure engineering approach needs to be
transformed to include a process which considers at its
core, the aesthetics, usability, and the end user experi-
ence of these new devices and their UIs. This holistic
approach will need to ensure that the disciplines which
span Design, Computer, Social, and the Physical sciences
are engaged within an interdisciplinary research and
development activity. Some early exploratory work from
various research groups have begun to show the potential
of such an approach over traditional engineering dominated
methods. As this work matures, a continual reflection of
the benefits of these new mobile devices should become part
of the development cycle to ensure that the maximum benefit
and an enhanced user experience is achieved.

Sam Bucolo
```

Sam Bucolo is Research and Development Director in Australasian CRC for Interaction Design (ACID), and Associate Professor in Industrial Design, School of Design, QUT

6.4.1 Common UI Design Guidelines

Before introducing the design process, we review a few important principles that should affect the UI design. A myriad of "UI design principles" exists; perhaps the most famous ones are those by Nielsen (1994) and Shneiderman (1998). These principles form the cornerstone of a good UI in general, but they are even more important in mobile devices, as the users have no opportunity to read a manual when they are on the move. Next, we summarize our UI rules.

6.4.1.1 Keep It Simple!

Among all the important design principles, simplicity is the one above all others. Use common sense and try and avoid any complex and "exotic" design solutions. To exaggerate a little, the world is full of wonderful inventions that are too difficult to use, for everyone except the designer. Fancy designs often demand a lot of labour during different phases in the product lifecycle: when being designed, implemented, tested, and – finally – when the users start to learn to use the system. There is, in most cases, no comparison to the gained benefit.

Instead, the best way to create usable user interfaces is to compare different solutions – traditional and radical, improve, innovate, test and iterate the design, listen to the users' voice, and – use loads of common sense.

A simple design should make it predictable what can be done now or what to do next, as the actions form a logical sequence with clear start and end points. When things are visible in the user interface, the user's cognitive and short-term memory load are reduced. The user should not have to remember what was on the previous screen; instead the system should provide the necessary information and cues. Also, the design should contain aspects that are familiar to the user as well as being easy and quick to learn.

6.4.1.2 User Controls Everything!

Users – especially those with a lot of experience – wish to feel in control. Therefore, the system should provide timely feedback on the user's actions, so that they know what they are related to. For frequent actions, feedback should not disturb the task flow. A rule of thumb is that feedback is considered instant if the system has provided a response within 0.1 seconds.

One second is considered to be the limit for the user's flow of thoughts to remain undisturbed, even though the user will notice a delay of such a length. Beyond this, the system should provide some indication about the delay and ten seconds is the maximum to keep the user's attention focused on the current task. For longer delays, the user can perform other tasks while the process is going on.

Feedback is also something that users appreciate when something sudden and unexpected happens, which leads us to the next rule. This also relates to dealing with external interruptions that occur frequently in mobile use, such as an incoming call while taking a photo. In such cases, control needs to be returned to the user as soon as the interruption has been taken care of.

6.4.1.3 Prevent Errors or Aid in Recovering from Them

It is unlikely that we can design a system that will prevent the user from making any mistakes. Thus, we should try and minimize the number of errors and make recovering from one as simple as possible. Errors can be small or large and caused by various reasons, such as misreading, misunderstanding, too much haste, and so forth. A small example of error prevention is that an application shows the desired time format when the user needs to input time.

If an error occurs, the system should provide constructive feedback about what went wrong and how to recover. Most likely one of the most frequently used actions is "Undo", which allows reversing from the current action to the previous one. Thus an opportunity is provided to step back, thus relieving anxiety and encouraging the user to explore the application.

6.4.1.4 Efficiency of Use

Skills and knowledge that the user already has should be fully used, since they promote familiarity and consistency. Consistency relates to several aspects of the UI, such as graphics, navigation, interaction, terminology, and component layout. It implies that the UI should look and behave in the same way, regardless of the task. Also, consistency between applications that use the same platform is important, allowing the user to apply things that they have learned earlier.

Since applications are used by people in different stages of the learning process, with different kinds of experience, we should not force them to change their habits, but instead to support alternative working patterns. A novice user requires more navigational steps and aids when performing a task, while an experienced user completes it routinely by using only a few steps. For example, in Series 60 devices, a novice user needs an explicit method to select multiple objects on a list, by using the "Mark" command in the menu. For experienced users, there is a keyboard shortcut available, by pressing the dedicated hardware key while selecting objects. Also, creating an efficient tool requires the designer to understand that the users have often developed their own style for working. Therefore, it is beneficial to support UI customization and personalization.

6.4.1.5 Aid the User

Aid to the user is important, since they may still be in the learning phase, and not yet aware of all available possibilities, or the task may have been interrupted for a while and the user is returning to it. Aiding the user also minimizes cognitive load. Therefore, the UI should communicate its current state, especially if the state limits the availability of features or navigation steps, such as in an edit mode, where the user is only able to change the values of the object's attributes instead of using the object.

As far as mobile devices are concerned, online help is a primary source of assistance, since it is unlikely that the user has any other additional help material available. It is expected that the user will need to find information about some actions, also when not trying to recover from an error. To make this easier, the system should know or predict what the user wishes to accomplish, based on, for instance, the state of the system and available objects.

The system should direct the user with clues and hints on what are possible and typical actions, and how they could be performed faster. For example, the shortcut key shown next to the menu item's label educates the user to use an accelerator; bolding the default menu item suggests the most typical command; and disabling items shows what cannot be done now. The order of actions in a menu or toolbar may represent the frequency of use and so frequent commands require less navigation steps. Similarly, the layout can be used to emphasize importance, or the order of filling in information.

If the order of filling is related to critical and rare tasks, the UI may contain a *wizard*, which divides the task into a set of steps and associated views and controls, which the user must perform in a specific order. The wizard structures the task and assists the user to perform actions in the correct order, and to input required information by showing only the controls and information that are relevant to each step. Obviously, the wizard is not flexible, as it controls the task flow instead of the user. Consequently, wizards should be used sparingly in the UI.

Because text input with mobile devices is slower than with desktop computers, the UI designer should think of ways to reduce it. A technique referred to as *auto-fill* tries to predict what data the user wishes to input in a certain field based on, for instance, the system's and the application's states, the type and the format of the input field, the previous values of the input field (history), and user preferences. Another way is to show whether the field is mandatory or optional; in the case of an optional field, the user may skip it.

To save screen space, additional information related to an object may be hidden until the user explicitly requests it. A popular method is known as a *tooltip*, which provides textual, and sometimes graphical, information for an object as a textbox superimposed over the object. The tooltip is displayed after a certain period of time when the user highlights the object, or the user hovers the pointer over it.

6.4.1.6 Create Aesthetically Pleasing Designs

Even though it is essential to design for efficient, flexible, fast, and error-free UI, it is also important to create elegant, stylish, and aesthetical designs. This often leads to a situation where the designer must make a decision between enjoyable visual design and efficiency. For example, selecting a stylish colour palette can easily result in reduced legibility of the text due to poor contrast, or using graphical dividers between screen sections overcrowds the screen and slows visual searching.

What is considered beautiful is subjective and culturally dependant. Therefore, the UI should be customizable and personalizable. For example, the user can change to a more familiar set of icons, and change the order of objects.

Minimal visual style is often considered beneficial as the information that is rarely needed or irrelevant to the task is hidden. Also, because the UI contains little perceivable information, the user's attention is guided towards parts that are the most important.

6.4.2 The UI Design Process and Methods

Now we are ready to illustrate our approach to designing mobile user interfaces.[13] A mobile UI is an aggregate of many factors that should be taken into account in the design. For example, Lindholm et al. (2003) characterize *user interface style* as: "A combination of the user interaction conventions, audio-visual and tactile appearance, and user interface hardware."

The design of the system involves many concurrent activities. UI should be considered in the early phase of the design, when analyzing

---

[13] Our UI design process is influenced by the OMT++ method (for further information, see Jaaksi (1995)). We do not claim that our process and method is the one and only – instead we wish to illustrate the phases of the user interface design.

requirements to make the functionality and technology accessible. A mobile device is often a single unit, therefore, different kinds of form factors and interaction device configurations will have an impact on industrial and hardware design, etc.

Regardless of methodology, the UI design process consists of four distinguished phases:

- analyzing users, tasks, and operations
- specifying UI structure
- defining UI components, and
- layout and visual design.

In the first phase, the target users and their characteristics (such as background, experience, and language) are defined, and the tasks that the user should be able to perform with the application are analyzed. The designer must understand why the application (for example, a music player) is being designed and what are the most important tasks (such as playing back a playlist, creating a playlist, or browsing the music library) that the user wishes to accomplish. Often the tasks are presented by use cases or scenarios, which are spoken language descriptions of ways to use the system, related sequences of operations, and the operating environment.

Based on the task analysis, the designer can start to identify the operations that are required to perform the tasks. Operations are elementary procedures consisting of a series of actions that the user performs. Actions are the smallest activities that the user can perform with the actual user interface (for instance, pushing a button, or scrolling a list). Making a playlist requires the user to create a new playlist, to select desired songs, and to save the playlist.

After the operation analysis is complete, the designer summarizes all the defined operations as a list to identify what unique operations the tasks share. This way the designer can keep the design simple and consistent between the tasks.

Before moving on to the next phase, the designer should be aware of the platform and other technical aspects, such as technical constraints, requirements, and possibilities, which will influence the design. For example: What kind of form factor is used? What is the display size and resolution? What kinds of input devices are available for the interaction?

Furthermore, the platform may provide an UI guideline (Series60. com 2005; UIQ Technology 2002) for the designer that describes the practices and conventions used in the platform. The guidelines, if used properly, support learning and understanding on how the system can be used, as they aid in creating a consistent design. Also standards, such as the ISO 9241 series, offer guidance by providing information, requirements, and recommendations related to certain hardware or software (ISO 1998).

After sorting out the technical issues, the design continues. In order to guide the design for the UI structure and component specification, the designer should consider the following questions and try to answer them during different phases of the design (Aaltonen 2007).

What content should be presented? This first question is fundamental, since it sets the boundaries for all subsequent questions. The answer depends on several factors, such as the current context, current user task, previous states of the UI, user preferences, and device capabilities. The amount and type(s) of content should be defined, because this will set the frames for the other questions.

How to present the chosen content? After knowing what is going to be presented, we can select a suitable presentation method. The previous system state, selected content, system hardware and software capabilities, and user preferences, etc., have an effect on the presentation. Furthermore, presentation and interaction, addressed below, are heavily interlinked. To define a presentation technique, we should first select the output modality, such as speech or visual. The modality used in the previous state should usually remain, because it makes no sense to switch modalities frequently. Then again, synthesized speech is troublesome for presenting dozens of images, as the output modality does not support the content modality. Alternatively, speech is fine for synthesizing the contents of a text message.

The presentation technique depends on content type, but we need to consider what information is shown for each object (what metadata attributes and their values are presented and so forth) and how the information is presented: as text, graphics, or audio. If the visual presentation is used, the screen resolution, orientation, and size affect the layout to a great extent.

How to interact with the content objects and information? The objects and presentation guide interaction, which depends on the previous system state and selected objects, as well as system hardware and software capabilities, user preferences, and so on. The presentation technique is tightly linked with the input capabilities of the device, because the presentation should be such that the user can interact with it. Also, the interaction occurs on several levels – with the highlighted object, with the current view, or with the application – and how to design the interaction is affected by the available input devices. Also, the current context affects the presentation and interaction: if the user is walking, they cannot manipulate a long list easily, while using a speaker and text synthesis would not be socially acceptable in a theatre.

How to provide and access functionality? In addition to implying what are the available actions or content object(s) relevant to the task, the question relates to aiding navigation between the objects and controls. How to perform an action or operation depends on many factors. If the action is performed frequently or the designer wants to highlight some core functionality, a hardware key can be assigned instead of using a menu. In essence, the different levels of functionality are:

- What operations are available for the highlighted object in the view?

- How the objects in the view can be organized (for example, ordered, filtered, or highlighted)? Can the type of the view be changed (from list to grid)?

- What application level operations are available? For instance, search, settings, shortcut to another application (go from camera to image gallery).

- Current state of the system (change WLAN access point).

How to navigate between the UI parts? Considering navigation in a single view or even in one application is not sufficient, since the user may need to perform multiple tasks at the same time and thus switch between the active applications.

The defined unique operations are required for designing the dialogue and the structure of the application. The aim of the structure specification is to place operations into states and depict the relations, or required navigation steps, between them. It is essential to provide fast and easy access to primary tasks(s) and related operations and thus, the main operations should be performed in the primary (main) state. Secondary operations, which support performing the main task(s) or are performed less frequently, are divided into secondary states. The structure of the user interface can be visualized with a dialogue diagram that presents the states and related operations and transitions (relations) between the states.

In order to illustrate the design process, we now consider a fictional dedicated music player device for sporty users. The device is strictly

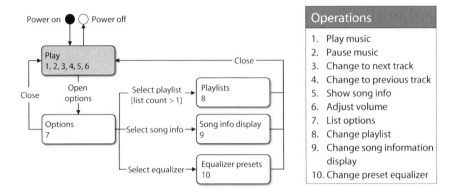

Figure 6-25. An example of music player's dialogue diagram and operations.

targeted at enjoying music and therefore, all remaining GEMS actions are handled via a PC when the device is connected to it. The design should also aim at creating a small device, where the number of mechanical interaction components is minimized to create a splash-proof product. Consequently, interaction is handled via a touch sensitive display capable of showing different shades of gray. The battery level of the device is indicated with an LED.

Figure 6-25 illustrates the UI structure of such a device with a dialogue diagram. The device has five states (views): play, options, playlist, equalizer, and song information. The operations related to the device states are presented with the related number in the diagram. Notation used in Figure 6-25 is for illustration purposes only.

In the diagram, "Play", which is the main state, is depicted with a thick border and grey background, while the white round rectangles present temporary states (modal dialogues). Turning on the power takes the device to the Play state. To change the current playlist, adjust the way the song information is displayed, or access the preset equalizer, the user needs to first move to the "Options" state. In the diagram, access to the "Playlist" is conditional, indicated with the condition inside the brackets. Lastly, the power is switched off, only while in the Play state.

The purpose of this dialogue diagram is to depict the required structure and transitions. The designer recognizes what operations are performed in each state and consequently defines what kind of UI components are related to it. Furthermore, how the transitions from one state to another are done is defined. In mobile UI design, this phase is one of the most essential; if the structure design is poor, the user does not find the desired operations easily. Navigating between

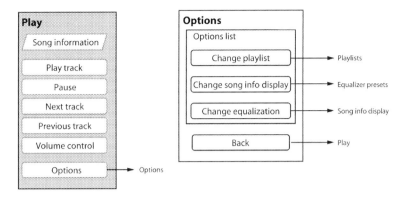

Figure 6-26. An example of component specification of a simple music player.

the states may then require too many navigation steps that hinder its use.

The next phase is known as component specification. Here the designer identifies what controls are required for performing the operations, and how the transitions to other states are initiated. Furthermore, in this phase, UI guidelines identify whether the platform provides the desired UI components or performs the design-required custom components. Figure 6-26 depicts the component specification of two of the states for the music player, Play and Options.

In Figure 6-26, the components for the main state are inside the grey rectangle, while the secondary states are within the white box. The UI components that can receive user input are presented with round-cornered rectangles, the pure presentation components are parallelograms, and the combined components are grouped together within a rectangle. If a component causes a transition to another state, it is shown with an arrow and label depicting the name of the subsequent state. This illustration is not sufficient for real life implementation purposes, but must be accompanied with detailed descriptions for each component that illustrate exception handling, valid data types and their ranges, default values, and so forth.

The last phase of the design is creating visual appearance with the aid of possible style guides for the platform. Even though the design follows the guidelines of the platform strictly, there is always room for aesthetic improvements and styling.

In addition to guidelines, it is important to notice that our example – the music player – is full of *de facto* standards and conventions. For example, the symbols that are used for presenting playback (▶), pause (‖), next track (▶▶), previous track (◀◀), and volume (◢), are well known

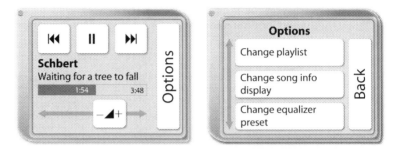

Figure 6-27. Detailed UI designs for the Play and the Options states for the music player.

globally. This also relates to the way the controls behave. For example, it is common that pressing the play button toggles between play and pause modes, and a long press of ◄◄ or ►► buttons will perform rewind and fast forward functions.

Figure 6-27 illustrates the detailed designs of the Play and Options states. The purpose of the design is to demonstrate the transition from the component specification to an UI sketch. We should be aware of the fact that there are several unaddressed issues, such as the lack of navigation aids, in this example.

Visual design is a significant part of the detailed design phase, since the final look and the design layout are being finalized. For further information on visual design, see Mullet and Sano (1995). When crafting the layout based on the component specification, one crucial aspect is to find the right balance between showing enough information and overcrowding the screen.

Another aspect is to consider the most important information and the controls that are most frequently needed. Failure to understand the tasks and operations easily leads to information overflow and complex menu structures. Furthermore, to create a layout that is easy to scan, the design should consider reading orientation and alignment of UI components, which guide visual search.

Icons often have a central role in graphical user interfaces, as objects or commands are associated with them. Therefore, good icon design focuses on issues such as identification, expression, and communication.

6.4.3 Validating the Design

A term often associated with user interfaces is usability (also referred to as ease-of-use), which is defined as: "The extent to which a product can be used by specified users to achieve specified goals with effectiveness, efficiency, and satisfaction in a specified context of use." (ISO 1998).

The user interface should be easy to use, but how can we say that one UI is easier to use than the next? We need to evaluate the UI and obtain results that can be measured. In this section, we list some of the measurable usability factors (Dix et al. 2003, Nielsen 1994; Shneiderman 1998).

Learnability refers to how long it takes a novice user to learn the commands required for performing a set of tasks. Often this is presented graphically in the form of a learning curve, where the x-axis shows time and the y-axis presents the number of mastered tasks or an equivalent measure. By observing and comparing learning curves, one can claim that an application is easier to learn than another.

Learnability is difficult to measure in a short test session. In this case, we can say that we do not measure the learning process as such, but *immediate usability*; for example, something that is related to how logical the structure and navigation is; or how familiar the terminology is. Also, if we perform a series of tests with the same application and users, we can study how well the users have maintained the knowledge and skills they have learned over time.

Effectiveness relates to how well the user can achieve the goals, measured by the percentage of accomplished goals or the number of features used, to name but two criteria.

Efficiency determines how long it takes to perform a certain task. Typically, these are the most fundamental and most frequently performed tasks. This is a useful measure and easy to do since we can measure time (typically milliseconds) provided that we have defined the start and end points of the task in question. However, we do not necessarily have to count seconds; instead, we can count the number of mouse clicks, key presses, or navigation steps.

How many and what kind of errors do users make when they are performing frequent tasks? It is error rate that we are referring to here. Counting errors and removing their sources from the design is important for user satisfaction. For example, if you have a speech recognition system with 90% accuracy, it still means that every tenth word is recognized incorrectly and the user must react accordingly. Also, it may be worthwhile to study how long it takes to recover from the error.

Subjective satisfaction is how pleasant the user perceives the design, depending on the attributes above, as well as visual and auditory design and styling.

There are different ways for evaluating usability. Common methods include heuristic evaluations, where the experts review the application based on certain guidelines (heuristics), point out possible problems, and suggest improvements; and usability testing, where the test subjects perform a set of test tasks with the application while being observed. There are also a myriad of additional methods that are useful and important. It is crucial to understand that often selecting a usability test method does not rule out using other techniques.

Furthermore, the designers should set usability metrics, where they describe the target level for the usability in terms of desired values for the above-mentioned attributes. In addition, the worst and the best case values should also be defined.

6.5   Performing the GEMS Tasks

We have illustrated various aspects related to interacting with the device in general. Now we discuss how they relate to various GEMS tasks. Furthermore, we describe what kinds of steps are required in the tasks, and how to use different views and controls in order to aid the user in reaching a desired goal.

6.5.1   Cross-GEMS Tasks: Browse and Search

Before performing any GEMS actions on a content object, it must be first located. There are two ways to do this: either browse or search. Which one to use depends on whether the user knows where the desired content object is stored and how to access it. If the user knows the location, they can *browse* (or perform implicit search), otherwise they need to (explicitly) *search* for the object. However, because the search result set typically contains more than one result, the user still needs to browse. These two are frequent tasks that need to be performed before executing tasks at later phases.

If the user cannot explicitly name or locate the desired item, they must start searching.[14] Searching is performed either in real time, or it is made against a pre-indexed database. In addition, some systems, such as peer-to-peer networks, do not maintain directories of the content they contain, which means that the user cannot browse the content without searching first.

---

[14] But then again, Google desktop search is used for accessing local content instead of browsing, even though the location and the name of the desired file is known. Typing a (partial) filename in the text box is considered faster than navigating in a file browser.

Searching, as a human process, was introduced in section 6.1; a number of searching techniques were introduced in Chapter 5; and how metadata aids searching was explained in Chapter 4. Now we focus on the user interface: formulating the search query (search criteria) and displaying the search results (selection criteria).

Often searching is based on finding keywords (textual data) from content or associated metadata. In some cases the search may be extended to exploring text that is related to the (embedded) content object. For instance, to find a certain image, the system looks for words that are part of a message containing an image, or words in the paragraphs above and below the image in a document. Consequently, the simplest search user interface is a single textbox, where the user can type the query.

If the system supports re-using previous queries (which supports human search behaviour, as discussed in section 6.1), the component is a dropdown list box (Figure 6-18), where the user can select a previous search string and edit it before starting a new search. Also, the system can support saving searches and results for enabling further re-use of previous or frequent searches, and provide offline search result exploration. Obviously the user may enhance the search by adding advanced search criteria (Figure 6-18). The user can define values and value ranges for attributes, such as time, sender, content type, location, object size, date, or other metadata. Furthermore, it is beneficial that the user can prioritize the attributes, which means that the attribute of a high priority has a larger impact when calculating the relevancy of a result.

However, it is not necessary to enter keywords. The relevance feedback method (section 5.2), allows the user to select an object or part of it as a criteria for a search. The leftmost image in Figure 6-28 demonstrates this by showing a contextual pop-up menu of an image embedded in a message. After selecting "Find related" menu item, a view on the right image is displayed. The view consists of "the seed image" and list of related items. The common attribute values are bolded in under the seed image.

Instead of finding related information, another option would be to search for similar content. This differs from the previous method in such a way that the objects do not have any relations between them; instead they should contain similar features. These kinds of searching will become even more powerful as technologies for recognizing features (such as human faces in an image, video footage shot outdoors, or recognizing the song genre automatically) of the non-textual content evolve, thus introducing new search attributes.

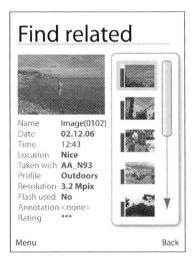

Figure 6-28. Schematic UI for searching for related content.

Furthermore, an interesting aspect in digital searching is collaboration. Information retrieval researchers have suggested that information seeking has always been a social process (Wilson 1981). Collaborative searching concerns search query generation, result browsing, or both. One form of collaborative searching is *collaborative filtering* (Maltz and Ehrlich 1995). Collaborative filtering matches other users' interactions with the user's own interactions to facilitate the search process. In order to function properly, collaborative filtering requires a critical mass of users.

Meta-search is a proposal for implicit collaborative searching in peer-to-peer networks (Lehikoinen et al. 2006). In addition to performing regular searches, this method supports searches based on other network users' previous searches on the same or similar topic. In essence, when a user performs a search, they receive not only the usual result set, but also information on other users' previous results, as well as relevant information (such as how many times a resource that appeared in the result set was successfully downloaded). This way, the community provides support for searching without any other extra effort; the support is a side effect of the normal searching process.

While the search is ongoing, it is necessary to show some kind of a progress indicator to let the user see that the system is performing the operation. After the search is performed, the search results are presented and the task switched to browsing. The result set is typically ordered by the computed relevancy of the results. The system can exploit other user's behaviour, such as what items they opened or downloaded, with similar search result sets in order to fine-tune the computed relevant values (Lehikoinen et al. 2006).

In general, browsing can be considered a single task, but it is often the first step towards further GEMS operations. If the user knows the location of an object, they can simply move through a navigation path to access it. The navigation path depends on the object's location, for instance, whether the object is stored locally or remotely, and how the objects are organized. In browsing, the fundamental design decision deals with defining what objects are presented to the user in some meaningful way that best supports the user's goal.

What is required for the UI in content browsing? First, we need to select the objects that are presented to the user. In file management, this means showing the content of a folder or folders. However, with the power of the metadata we are not tied to folder structure, but free to use more sophisticated methods for this purpose, such as associations, relations, and dynamic user-defined queries.

The selection of the presentation modality and technique, as well as interaction technique, is tightly coupled. In addition, it is imperative to consider various factors that affect their selection, in order to provide a best possible solution that is beneficial to the user's current task and context.

Earlier, we discussed list- and grid-based presentation. Now, we introduce a few concepts and discuss two fundamental issues related to browsing: maximizing the amount of the content displayed on the screen, and modifying the view.

Media Tray (Hakala et al. 2005b) photo browsing application attempts to maximize the amount of content displayed on the screen, by hiding and making transparent the UI controls that are not required. Furthermore, it compresses the content into a more condensed form, by slanting the view and overlapping the objects slightly (Figure 6-29).

During browsing, it may be necessary to change the way the objects are presented in the view. This may include changing the way the content is organized, or even changing the presentation technique. As the Media Tray is designed for personal visual content, organizing the content based on time is important as the various events where the content is captured are related to certain dates. Consequently, the application provides a time bar at the top of the screen to filter and control the time span and thus the amount of content on the tray. The time bar consists of two rows; the first row presents the time span (a day, week, month, or year) within which images on the tray area are displayed, and the second row will change according to the first row selection. For instance, if "Month" is selected, the second row shows the month names (Hakala et al. 2005b).

Figure 6-29. Time bar (on top) and slanted tray of the Media Tray application.

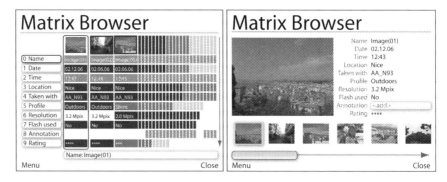

Figure 6-30. Matrix browser provides tabular and list views for interacting with metadata.

Because the amount of metadata associated with objects is considerable, it may be beneficial to visualize metadata attributes and their values. Matrix Browser (Figure 6-30) applies Bertin's Reordable Matrix for providing an interactive visualization of a set of objects and their metadata (Bertin 1981). The metadata is displayed in a table, where each object forms a column and the metadata attributes are then mapped along the rows. Studies show that the original Reordable Matrix can be used without training and the users are able to make valid observations from the data (dependencies) and realize how the interaction with the matrix works (Siirtola 1999).

In the matrix, the attribute names are displayed as row headers. In each row, the cell values are presented with the aid of colour coding ranging from dark blue (the highest) to pale blue (the lowest). Grey indicates attributes with highly volatile values, and white represents a non-existing or missing value.

As the number of objects that needs to be presented is often high, the columns are narrow so as to display as many objects as possible. The currently highlighted object and the two subsequent objects form a so-called *lens*, which shows the items enlarged in order to reveal more details, while the rest are displayed as lines only. The cell value of a selected attribute and the object is displayed in the message area below the table.

In addition to the table view, Matrix Browser provides a more traditional list view for viewing the current object and its associated metadata in detail. Furthermore, the user may select an attribute and edit its value in the list view. The user manipulates the matrix by changing the order of the rows in order to indicate the importance of an attribute (the topmost row is the most important). Furthermore, the user can sort the columns in descending order by using the attributes as a sort key, which makes the leftmost column the best result.

Matrix manipulation lets the user explore the data and find dark blue areas in the matrix, in order to discover attributes with mutual dependencies. Row selection may be toggled on and off by using, for instance, a number key on the keypad, associated with a row number. The location of selected row(s) can then be changed with up and down cursor keys. The lens is moved with left and right cursor keys and the columns are sorted to descending order with the "*" key; finally, the "#" switches between the views.

Matrix Browser can also be used for other purposes, for instance, visualizing a result set of a web search query that is based on several search terms. Then the rows will present the search terms and columns result pages. Each cell will then display how many times the search term has been encountered on the page. The darker the colour, the more hits on the page.

As Matrix Browser and Media Tray illustrate, it is important to allow the user to interact with the view, for instance, to filter out irrelevant content, or to change the sort keys and order. The user should be able to change the presentation technique, for instance, in a map-related application it may be useful to browse all the points of interest (POIs) as an alphabetical list, but it may also be useful to see them overlaid on a map (personal content and location information is discussed in detail in section 7.2).

Metadata is often considered additional information that should be hidden from the user, unless they explicitly wish to view it. Nevertheless, there are occasions when the metadata attribute and its value should be displayed and it can also be available for editing attributes such as date, time, keyword, rating, annotation, and comment. If visible, the metadata is displayed mainly as a text string (such as "creator: John"), but occasionally it is visualized (for example, "rating:★★★").

6.5.2 Get Content

In the GEMS model, the "G" phase is related to acquiring new or existing content object to the user's possession in some way. Next, we discuss downloading, where the user actively and explicitly initiates getting an existing object; and receiving, where the user's role is more passive (they get content without any interaction with the device), and creating new content object as an essential way to actively get content. Because the phase is at the beginning of the GEMS model, a question related to each task in the category is that upon getting a content object, where is the content stored and how can the user access it later? For instance, currently the Series 60 has application-specific download folders in the file system, which means that new content is distributed across several locations. Furthermore, if the folders are not empty, new objects are added to the existing collection(s).

Download is a task that is strongly related to searching and browsing, as they are often the preceding tasks. In essence, the user desires to get an object that is not stored in the device, which implies that a transfer is required. Initiating a download is often simple; just point and click a control, such as a button, link, or icon, and the process of transferring the object to the local device can begin.

Sometimes the system presents additional options (i.e., store or open the object after download) with the aid of a dialogue window. Once the download is started, the system should present the task progress, including an estimated time of completion. It is also important to provide UI components for controlling the process, as the user may wish to pause or cancel the download. Figure 6-31 presents an example UI for managing a download processes.

Multiple simultaneous downloads must be supported. Figure 6-31 shows a list consisting of four downloads at different stages; the first one is completed, the second and the third are ongoing, and the last one is in a paused state. For each download, the list item shows the object name, format, size, colour-coded progress indicator for transfer, cost indicator, icon for the transfer channel, state of transfer, estimated amount of remaining data, and time. Furthermore, each item has controls for interacting with the item, their functionality depending on the item state.

There are many technologies for downloading content. Typically, the technologies have a different effect on the cost and duration of the download process. Because users are dominantly cost-aware, it would

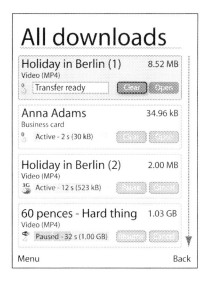

Figure 6-31. An example of UI controls for managing downloads.

be beneficial to display some kind of estimate of how much the download will cost. In addition to various technologies, this becomes more difficult due to numerous pricing logics: free of charge, flat rate (monthly fee), based on transfer duration, and based on amount of transferred data, pay per use, and pay per object.

Even though the cost is difficult to estimate, the system should provide information on how long the download will take. This is significant, since the user may be mobile and they should have some indication of when the content will be available for further operations. In Figure 6-31, the list item shows the price group with dollar signs, which aids the user in estimating the cost of transfer.

Receiving an object is a passive task from the user's perspective, because they do not initiate the task: the user receives some content, even though not requested, whereas when downloading, they have explicitly expressed that they desire to get the content.

There are numerous wireless technologies for receiving content, such as infrared, Bluetooth, text messages (SMS), multimedia messages (MMS), and e-mail (with attachments). More recent technologies that could be used for receiving content include RFID[15] and

---

[15] RFID (Radio Frequency IDentfication) refers to technology that resembles a barcode, but identifies items with the aid of radio waves. A reader communicates with a small tag that stores a digital identification number.

NFC.[16] In addition to RFID, barcodes can be used as a links to a certain service and receiving content indirectly. The rapid spread of camera-equipped mobile phones has boosted the use of cameras as visual scanners for reading the code.

Mobile devices consider an object received by any of the afore-mentioned technologies as a message, and store it in the messaging inbox. The type of message is then illustrated in the inbox with an icon. The metadata should be used for organizing the inbox, and thus enabling the user to interact with the view: filter off content; change sort keys and order; and group the content. The user may select objects for further manipulation, which then leads to other GEMS phases: enjoying (viewing, listening, editing); maintaining (organizing, rating); or sharing (forward a message).

Since the user did not initiate the transfer, they must be notified of its arrival. The notification should provide a direct link to the object, so that the user can access it quickly. However, the user may wish to discard the notification because of the current use context or task. If they decide to do so, the system can provide a smaller, unobtrusive peripheral indicator that reminds the user of the received item.

Depending on the type of category, current mobile devices support *creation* of a wide range of personal content objects, such as contacts, documents, drawings, images, messages, notes, points-of-interests, spread sheets, videos, and so forth. Obviously, creating different kinds of content objects in various formats requires different kinds of views and controls.

Creating content is often one of the most frequently performed tasks with the mobile device and may be one of its main selling points. Therefore, the user interface for this task is extremely important.

Since the creation task does not use metadata directly as a user benefit (since no metadata exists yet), we do not focus on it profoundly. Instead, we wish to emphasize that the creation is a truly "once in a lifetime situation" in the GEMS model and therefore is extremely important. In addition, to enable further (and perhaps even automate some of) content management tasks, the context information that is available at the time of its creation should be captured and stored as a part of the object's metadata.

Metadata can also be added as pre-defined *keyword* tags – such as holiday, mum, dad, Xmas – that are associated with the object. The user does not have to edit the keywords; for instance, merely drag and

---

[16] NFC (Near Field Communication) has evolved from RFID and is a short range (few centimetres) two-way wireless connectivity. It allows reading small amounts of data from other devices or tags.

drop the suitable tags on top of the object. Even though the user does not edit the tag's keyword, the fact that the user has explicitly selected the tag implies importance.

Shooting images and videos is one of the most popular tasks for getting content with a mobile device. In a simply form, capturing an image or a video task requires a viewfinder and zoom, flash, and capture controls. Occasionally, the user may wish to override automatic settings and, for instance, modify white balance and exposure.

Another method to get visual information is *scanning*, which changes content from a physical to a digital form. It may involve optical character recognition (OCR), where the system identifies characters from the image and converts them to text.

Recording audio is a less frequently used creation method and mainly used for dictation and note taking purposes, and as a secondary modality when shooting video.

Writing a document, composing a message, or listing to-do items requires lots of text entry. Furthermore, writing tasks may include embedding other content objects in that text. Text creation implies a need for formatting and styling the text, and supporting formats such as rich text format (RTF). Typically, the operations include changing font faces, sizes and colour, style, and formatting paragraphs (alignment and indentation, for instance). To enhance the presentation, the user may wish to add special characters and symbols.

Increased (informal) text-based communication, first via IRC and later via chat, instant messaging, and text messages, has introduced ways to decrease the number of characters that the user needs to type and enriched the communication with new terms, abbreviations, smileys, and emoticons. The older popular terms like BR, IMHO, and 2nite are now have now been added to by dozens of newer expressions, such as G2CU (good to see you), ILBL8 (I'll be late), and 404 (I don't know), which are challenging for predictive text entry solutions, not least due to their dynamic nature. A smiley is used to express the mood of the writer with ASCII characters. For instance, :-) or :) means happy, :-(is sad. An emoticon is essentially a way to present a smiley in a more graphical and perhaps personal way.

Drawing is among other tasks useful in making quick sketches. It benefits from a direct pointing device, such as a stylus on a touch sensitive screen, in comparison to an indirect device. Drawing predefined shapes, such as a line, an ellipse, a rectangle, or a polygon is doable with both input device types, but free-form shapes are difficult to draw with a joystick. Furthermore, changing the drawing tool (from pen to eraser, for instance) may be tedious with an indirect pointing device, as it is slower than with a stylus. A drawing can also be seen as a part of editing; for instance, the user may wish to emphasize a region of an image or add some text.

Related to creation (as well as editing), the *clipboard operations* (copy, cut, and paste) are useful when writing and drawing as they aid in, for instance, duplicating or moving parts of the content easily. These are also examples of operations that are so frequently performed that any experienced user probably prefers to use shortcuts. Copy-paste operations also offer a nice example of adding metadata. If text snippets or parts of the image are pasted to a new content object, the ID of the source should be added as a part of the metadata in order to imply the relations between objects. In general, this relates to creation of any kind of object that is composed of (parts of) other objects (slide show, presentation, playlist, personal map).

Creating a new content object by *composing content objects* is a task that is either related to a single content type (creating a slideshow from a set of images, a song from a set of short audio tracks (loops), or a playlist from a set of songs) or it may relate to creating a multimedia presentation with text, drawings, images, and video. Nevertheless, composing content is a question of defining the desired set of existing objects and their spatial, and in some cases temporal, order. Composing requires searching and browsing when seeking for the objects to be included.

With the aid of metadata, the system may provide an *auto-summarizing* feature that searches for the best candidates for the content. The generated set often acts as a starting point for further manual composition. Figure 6-32 depicts an example of a view that is the result of auto-summarization. Instead of using a single content format, the summary is based on time (shown on top-right corner), consisting of a map, points-of-interest, and images captured in those places. The label of an image is based on location or derived from a message that has been sent on the spot.

In addition to the above-mentioned personal media, there are other personal content objects that the user may wish to create. For instance, creating a new contact or calendar event requires filling in textual information (name, contact information, address, birthday, and affiliation), as well as perhaps attaching another objects. Or the user may wish to create a personal ring tone by composing a MIDI song.

Another increasingly important example is *creating a point-of-interest* to a certain geographical location (POIs are discussed in more detail in section 7.2). The simplest way is to get the current coordinates and store them for later use, but it may be reasonable to attach some label that describes the place and implies its relevancy. The label could be textual, which will make the search easier, but it could also be an image

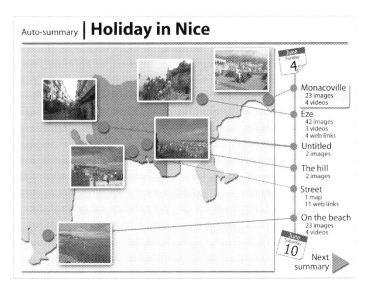

Figure 6-32. An example of multimedia presentation made with auto-summarization.

captured right on the spot that would facilitate visual searching and recall.

Another form of getting content is *ripping*, which has a bad reputation due to its illegal distribution of commercial content via peer-to-peer networks. Ripping is characterized as:

• converting or extracting data from one format to another; and/or

• copying physical media that is more useful to the user.

Typical examples include converting audio from a compact disc to MP3 files, or scenes from a DVD to MPEG4 files. The actual conversion process is straightforward from the UI perspective, not requiring any fancy controls. Since the task involves drives for physical media as well as lots of processing power and memory for transcoding, it is often done with desktop computers.

There is a form of ripping that requires more user involvement and aids from the system. This is related to, for instance, capturing images and text snippets from documents (such as WWW pages) and inserting them into other documents that are in a more convenient format (such as ASCII text). This kind of ripping requires the system to support clipboard functions (copy and paste).

Finally, getting new content leads always – immediately or later – to other phases of GEMS. An example of these tasks could be capturing an image and deleting it immediately if the capture had failed. Consequently, it is imperative to consider the UI structure and what are the next tasks, and in which order to design tasks flows that are meaningful and relevant to the user.

6.5.3 Enjoy Content

Enjoying can only start once the user has got the content object, for instance, after a completed download (although some download clients allow enjoying partial content while it is still being downloaded). If the content is already stored in the device, the user may access it by browsing or searching. It should be noted that the content, which is being enjoyed, may be sensitive and private. Therefore the user should always have absolute control over where, when, and how the enjoying occurs. The user interface for enjoying relies greatly on the object's type. In this section, we present only the main tasks.

Enjoying – as any use of – content should affect metadata as it implies importance and/or relevance of the object. A typical example is the "use count", such as the play count for a song: the song rating could be suggested by the system, based on the play count.

Playback is related to temporal media types, including music, audio books, video, animations (like Machinima as discussed in section 7.4, or Flash), slideshow, podcasts, and videocasts.[17] The most common controls are play (that often acts as a play/pause toggle), and move to the previous or next item (track, scene, or image). In some cases, the controls include functions for stopping the playback, starting to record content, and changing the volume or the media source (video, audio). Depending on the device category, the playback controls are either drawn on the screen, or the device has dedicated hardware keys (as in Apple iPod or Nokia N91 (Figure 6-5)).

Although watching a video or listening to a song requires similar controls (such as play, pause, stop, rewind, and fast forward) for manipulating the temporal dimension, the view they demand differs greatly due to the modalities they need. Mostly this relates to information that needs to be presented on the screen. For instance, while studying mobile TV broadcasting, Knoche et al. (2005) found that images should not be smaller than 168×126 pixels, because otherwise

---

[17] Podcast refers to audio files (typically in MP3 format) that are downloaded over the Internet. The files may form a show, as in radio, and the user can receive the latest episodes with the aid of an RSS feed. Videocast extends the content from audio to video (MP4).

the details become too small and the acceptability of content is decreased.

A set of images can form a slideshow, where the images are displayed in a specific order. The slideshow is usually based on images that the user chooses, but also metadata-based relations can be used to create a dynamic slideshow based on the selected attributes. The slideshow either uses playback controls, which means that the user can move between images manually (transition to the next or previous image in the set), or the transition is automatically based on a pre-set time interval.

Above, we mentioned a slideshow in relation to playback, however *viewing a single image* requires different kinds of controls, including those for navigating between the images in a set of images, and changing the orientation. Rotating an image is easy when the device is handheld, but if an external device is used, a fast method (a finger gesture or keypad shortcut, for instance) to rotate an image 90 degrees clockwise or counter clockwise is a necessity.

In mobile devices, visual content often needs to be scaled down to make it fit the screen. In this process, some details are lost because the number of visible pixels is less than the actual number of pixels. In order to view the details, the user must zoom in, which implies that the whole image is not visible anymore. To see the parts outside the current view, the user must be able to pan (scrolling) the view. There are mechanisms (such as earlier mentioned MiniMap) that can aid in scrolling the view.

Essentially, zooming may be used with any visual content, although it is less frequently used with temporal content. This is because during the playback the user cannot be sure if something relevant has been clipped due to the zoom.

Maps are a special case of images. Even though some users enjoy viewing maps as such, they are often used for *way-finding* or *locating tasks*. A technical difference between most "normal" images and maps is that while most images, photos included, are bitmaps, maps are often vector graphics that enable a smooth change of zoom level. Wayfinding requires the user to enter the start and end points (sometimes also other waypoints) for the route, which is then automatically calculated. The waypoints are pre-defined points of interest, selected locations on the map, or addresses. Furthermore, the preferred route is usually weighted based on points of interest, scenery, distance, or travel time. It is straightforward to overlay the route on a map image, which makes it easy to see the progress as well as anticipate events and places ahead. It is beneficial to change the zoom level to see either more details of environment, or a wider overview. Especially in car

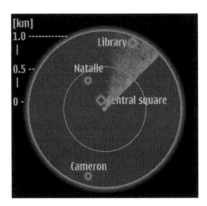

Figure 6-33. Displaying location information.

navigation, car speed dependant zooming is practical, since zooming out as the speed increases provides the driver with more time to react to forthcoming driving instructions.

Figure 6-33 presents an example of a visualization that shows the current locations of friends who are near the user. In order to minimize the amount of visual information on the screen, the visualization presents only the streets, the traveller's path, and two points of interest.

However, a map view is not necessarily required. Sometimes it suffices to present the directions as arrow indicators, and text or speech ("turn left after 200 m to Elm Street"). This is beneficial when the user is occupied with another task, such as following the navigational instructions while driving or riding a bike. This kind of minimalist information requires less attention.

Apparently all relevant POIs (as any content objects) may be difficult to display all at once while maintaining a proper level of detail. Halo is a technique for presenting objects that are off the view. For each off-screen object, a circle is drawn around it in such a way that it becomes visible on the border region of the view. The visible part of the circle implies to the user the location of the off-screen object at the centre of the circle (Baudisch and Rosenholz 2003).

Many types of personal content are in textual form, such as messages, documents, contacts, and web pages, which require reading. *Reading* a page requires controls partially similar to viewing images, but some additional controls for changing the font size (to improve legibility) and in-document navigation (between pages) are also needed. Some of the content is more interactive (web page, contact) than the others (note, message, document).

Figure 6-34. A sketch of social network visualization.

Visualizing social networks is an example of displaying textual information in graphical form. Essentially, it is about providing an overview to people, including the phonebook, e-mail contacts, IM buddies, and the like, in such a way that the user may see the strength of the social relations they form with others. The relations are based on several criteria: the call log, messaging history, common content, overlapping points of interest, Bluetooth encounters, and so on. Together this information can imply the strength and frequency of communication, collaboration, and shared interests between people. The social network, as networks in general, is often presented as a graph, where people represent nodes and the arcs between the nodes show the relationship. As the communication between people occurs in two directions – in and out – the arcs should display them both as, for example, two directed arrows. If the network is based on one person's contacts, the visualization results in a flat but wide tree, where the user represents the root node (Figure 6-34).

If multiple persons share their contacts, it is possible to create a true graph. However, in a mobile device and with the limited screen space, the interaction with it may be tedious.

Editing a content object is another form of enjoying. As a result of the editing task, the user either creates a completely new item or modifies an existing object. In essence, this leads to the need for version controlling.

Editing as a task depends heavily on the type and format of the content. For instance, even though scalable vector graphics (SVG) and

JPEG are formats for images, the actual editing is different, as SVG is vector graphics and JPEG is a bitmap. We acknowledge that editing different content objects require special and dedicated applications and technologies, but we do not review them further as they would probably be a topic of another book.

The edit task relates to many other GEMS tasks. For example, when composing a presentation, the user often wants to create a new object as a part of it. Most likely, the user also needs to browse and search for relevant content. Furthermore, to validate the end result, the user may wish to preview the set.

As the preceding example illustrates, editing is a task that produces a lot of relevant metadata. For instance, as the user combines several objects to create a presentation, a message, or a slideshow, this implies then that the objects have a relationship, which the user has explicitly created.

6.5.4 Maintain Content

Interacting with personal content involves increasingly using different devices that form a personal device ecosystem (Chapter 4). The fact that the content is distributed implies that it is of utmost importance that the user can *move, archive, synchronize, and back-up content between the different devices* (the technical problems related to such tasks are discussed in Chapter 5).

Currently, the desktop computer acts as a hub, where all mobile devices can then connect (from time to time). Compared to mobile devices, the desktop computer offers better computing, storage, and interaction capabilities. In addition, home PCs have often a fast, flat-rate broadband xDSL connection. These facts, together with other capabilities, offer a wide range of content experience features to all GEMS phases. For example, Lifeblog is a combination of PC and mobile software that is designed to store, organize, browse, search, and edit personal content files, such as images, videos, and messages that the users have collected with their mobile phones (Lifeblog).

Because maintaining tasks require exchanging information between devices and the amount of content may be massive (consider a full back-up of all personal content), the user appreciates a connection that is fast, cheap, and reliable. Besides, security needs to be considered as the data may be sensitive and private. The typical wireless connections for maintaining are WLAN and Bluetooth, which are convenient from the user perspective as a physical connection is not required. Other widely used methods are a USB cable, and a memory card for an *ad hoc* sneakernet.

Consider a user of a device of a toolbox category, such as a PDA or a smart phone, which has 8 GB memory. The device could contain thousands of personal content objects (messages, contacts, photos, videos, music, etc.), as well as applications and their settings that are of high personal importance and value and which are collected over the years and stored nowhere else. Most likely the content is not backed up or synchronized frequently, even though people acknowledge the fact that a mobile device is more likely to be damaged or lost than devices that are stationary and stored indoors.

There are many reasons for the lack of backups. First and foremost, no really usable layman's backup solutions exist at the time of writing. And even if they did, these boring tasks are considered unimportant – until the importance becomes apparent due to some unfortunate event resulting in information loss.

In a device ecosystem, the need for synchronization is frequent as the content is being enjoyed and created with various devices and – in some cases – by various people. What is important is to keep content up-to-date between devices. For instance, keeping calendar and contact information up-to-date requires synchronizing between a desktop computer and a mobile device – supposing that this information is dealt with on both platforms.

Archiving moves rarely-used content to a permanent, less frequently accessed storage, in order to clear storage space that is in active use. This means that the user wishes to save the content – after all, it is not deleted, but they are willing to pay the price that the content is no longer always accessible or at least accessing it is more expensive in terms of access time or dollars.

The tasks listed above should be frequent and highly automated, in order to be useful and to minimize the inconvenience caused by information loss. In addition, they should be run in the background in such a way that they do not disturb the user or prevent performing the primary tasks. The device should also provide a notification if there is a backup device and suitable connection available (and the device contents have not been backed up recently). The system should also provide an indication based on, for instance, the percentage of content that has not been backed up, instead of mere time. Furthermore, the system can inform the user of the possibility to auto-archive content that has not been used for a long time, in order to free some storage space. Once again, these notifications must be subtle and unobtrusive.

A frequently needed task that is a counterpart of creation is *deletion*, which is the definite endpoint of a content object lifecycle; especially, if the object has no copies, backups, duplicates, or other versions available. Deleting can be for several reasons, such as the created object is not of satisfying quality or is no longer needed. Because deleting an object is a final step, and can be done unintentionally, there are mechanisms that enable undoing the operation. Desktop computers have a special folder called a trash bin or an equivalent, which stores deleted objects and also restores them if needed. Because this folder consumes storage space it can be emptied (semi)automatically.

Compressing content is a method that speeds up any transfer processes, as well as saving storage space, if the system can provide enough memory and processing power to do it in real time. Otherwise it is advisable to do this operation while the device is in an idle state. The system can identify rarely-used content with the aid of metadata and compress it whenever in an idle state. Similarly, the system can anticipate user future actions based on their behaviour with the content objects and uncompress the content beforehand.

In Chapter 4 we claimed that *organizing* content is one of the tasks that profit from metadata, as it enables the user to interact with the view and change its contents dynamically. We presented a set of operations, such as grouping, filtering, and sorting, as examples.

The true challenge of creating dynamic views based on metadata is to provide view controls that enable the user to manipulate the view quickly and effortlessly. The dynamic view should enable the user to explore a huge content set in such a way that they can generate different kinds of views in order to, for instance, benefit of Gestalt laws and pick up new information, see an overview, identify clusters of related content, and so forth. Furthermore, the views should be formed in such a way that the user can understand what is presented and why. For example, what are the attributes used as a sorting key, and what attributes and range of values are used for filtering?

Metadata can be used automatically and dynamically to create views where the use of the screen space is based on the amount and type of content. The user can then modify, for instance, the sorting key, define what objects should be highlighted or filtered, or what attributes and values are displayed.

Matrix Browser (discussed earlier in this section) could be used for exploring metadata of content objects as well as an interface for modifying the values. However, in some cases, the user may wish to describe which attribute a certain part of the object relates to. For instance, the user emphasizes or selects a portion of text or image and attaches a keyword to it.

Even though metadata should be created largely through automation, there are a few tasks that require manual entry, especially *rating* and *annotating* the object. These tasks are related to the 'E' phase, as they should also be doable while enjoying the content.

The ratings can be automatic and based on use count, but the problem is that the value is a system-generated prediction, not a subjective one. Often the rating is presented on a scale from one to five stars, where one is the minimum and five is the maximum.

The user should be able to rate a content object while browsing or enjoying. For example, mobile devices with a numeric keypad support fast manual rating, as the user can rate the object with a single key press. Even if context information, from the time of object creation and throughout its whole lifetime, has been saved as metadata automatically, it still does not contain the user's own interpretation of the situation. Therefore, the possibility for user-created free-form annotations is important, as they can be used as keywords.

6.5.5 Share Content

A counterpart of getting is sharing, which relates to the need or desire to give or dispatch personal content to one or more people to facilitate communication or share experiences. With mobile devices capable of wireless communication, this relates to operations such as sending, publishing, and uploading. For other devices, this relates to presenting the object with the device and perhaps later transferring it to another device capable of electronic communication.

Personal content may be sensitive and private in many ways, which means that unintentional sharing of this kind of content (by automation, error, carelessness) should be prevented. There should be different privacy levels for content based on metadata by, for instance, applying user-defined keywords that make the content always private. Such content objects need to be shared only explicitly and to trusted parties.

It is also important that the user can be certain that the system will not share their metadata to others (preserve privacy or maintain anonymity) unless they explicitly wish it to do so.

Sending has a strong analogy to real-world sending; it requires that the sender knows the receiver – or at least their address (be it a phone number or an e-mail address) – in order to deliver the message. Furthermore, the sender must be sure that the receiver is capable of receiving what is being sent.

Sending a message, such as an e-mail, an SMS, and an MMS, is communication between people that are not in the same location and at the same time. The message is composed of a header (sender and delivery information, like on an envelope) and body part that contains the content. The message body often includes text, but it is not always necessary as the user may embed a content object in it (if the message format supports embedding).

A message with a content object and without text can also be considered as *giving*, because the user is giving something to someone else for their use and not necessarily communicating anything. Perhaps the technology for giving differs slightly from sending as it often happens via short-range wireless connections, such as RFID tag, Bluetooth, or infrared. Giving content can occur through sneakernet, which means that the content is shared through some physical media, such as a memory card, and often also in a face-to-face situation (although not necessary, as the content can be sent via ordinary mail). As a remark, even though its name implies otherwise, instant messaging (IM) is text-based conversation instead of messaging, since IM takes place in real time and is based on conversation rules (turn taking, for instance).

In any case, sending or giving requires that the sender has the receiver's contact information, knows that they can receive the content, and is able to compose the message. From the UI perspective, composing a message body may require text entry, and controls for embedding one or more new or existing content objects in the message.

Entering contact information is not necessarily straightforward. It is done either manually (text entry or optical character recognition from the business card with embedded camera), with the aid of a contact list (phonebook), or scanning the nearby environment to find wireless devices within the range. However, entering this information often determines the communication channel (a phone number cannot be used for e-mail), thus hindering other possibilities.

With the aid of interaction metadata and presence information, the system can introduce some "smartness" in this step. For instance, what is the preferred channel of messaging between the people, and what are the channels currently available?

Giving content via a (hot swap) memory card requires that the system indicates that the memory card is temporarily removed and therefore the related features (applications stored on it are removed from the grid or at least disabled) are not available; neither can it be used for storing content created with an application.

The aforementioned memory card giving example leads to another typical sharing task that involves people sharing the same location: *showing*. Experiencing personal content often includes social interaction between people, which means that sharing of content also occurs in a face-to-face situation, not only remotely. In such a situation it is customary that the audio-visual content is shown on the screen and viewed at the same time by others; alternatively, the mobile device travels from hand to hand in a ring.

In order to ease showing, some mobile devices have TV-out connectors for an external display. Also, there are accessories that are capable of receiving content via Bluetooth, and can be plugged into a TV set for easier viewing. Similarly, a headset plug connects the device to a home or car audio system, or the user may use an adapter to divide one plug into two. Or two users may share one head set.

Somewhere between showing and sharing is a *collaborative* approach, where the user can share their view in real-time over the network for others to see. What is displayed in the view depends on the active application; it may be a slideshow or a web browser. The user can also share their audio to others at the same time. Furthermore, if the control to the view is also shared, it is possible to edit content objects collaboratively.

Strictly speaking, *sharing* refers to a task where the user makes certain content objects available for download by others. Depending on the system and technologies in use, the user can grant access rights only to certain people and limit the number of simultaneous downloads.

So far we have discussed sharing content with the people that the user is related to in some way. *Publishing* is heavily associated with uploading content on the Internet, where the user typically has less control over who can access the content.

Publishing in a mobile blog is a good example of integration between mobile devices, applications, and Internet services. This seems to be an important evolution path in the mobile domain. For example, capture an image, launch a mobile device's web browser, navigate to the bookmarked web page, and upload the image on a blogosphere. Depending on connection speed, the size of the newly created image, and other factors, this chain of operations may take as little time as 30 seconds, after which the recent shot is available for millions of people to view.

We have discussed sharing digital content, but it is still important to be able to convert the content in a tangible, physical form that does not require any technical aids. Often this refers to *printing* of visual material, such as documents and images. However, printers are not integrated with mobile devices (at least not yet); they are an important part of the personal media ecosystem. Printing is done in different ways, depending on the available technology. For instance, the user may transfer the content with a memory card or over the Bluetooth connection to a PC. On the other hand, as discussed in section 4.9,

PictBridge standard allows printing an image directly from an imaging device. Nevertheless, metadata may be used for transferring the personal settings and preferences, as well as technical information related to printing.

6.5.6 Multi-Tasking in GEMS

The previous sections illustrated what is related to single tasks and some of the requirements they pose on the user interface. In real life the tasks seldom appear isolated from each other. Instead they form a flow where the user performs several tasks that belong to different phases of GEMS. For example, capturing an image, removing red eyes, and sending it instantaneously concerns the Get, Enjoy, and Share phases.

In addition, there may be multiple tasks ongoing at the same time; in this case, each of them belongs to a separate task flow. The following example illustrates this:

A REMARK YOU MADE

It's a City day again for Cathy. The train ride from Belleville to London is generally not very interesting, and it's pouring with rain (as usual), so not a lot to see outside the train. The friendly old fellow opposite her seems as if wishing to chat, but she would like to be on her own today.

Cathy puts on her earphones and listens to the Weather Report songs that Steve loaded into her phone. Cathy did not know how to use the player, but Steve insisted to "get her to the 21st century" and showed her how to access the MP3 player on her phone. Good; now she has a use for that function. Some time for herself.

Except that the bloody phone is ringing. Who dares to disturb that Zawinal solo? Cathy takes the phone from her purse and accepts the call. The music stops. It's Deena – she has lost her phone and would need to check tomorrow's appointment with the curator. Deena had sent a copy of the calendar event to Cathy and now asks her to check the time.

Cathy is bewildered by this task. She does not use the calendar, preferring her scribbled notes, but Deena keeps sending her those electronic appointments. Well, she could try, but how do you get to the calendar when there's a call on? Steve, help!

Cathy explains her problem to Deena, feeling lost. That old chap next to her appears to be listening, and finally raises his hand as if wishing to join the conversation. Cathy is a little annoyed at first, but that quickly turns to relief after the man explains how she can access all phone functions with that Menu button, even while speaking on the phone. That works! Cathy goes to the calendar and, after some fumbling, informs Deena that her meeting will be at 11:00 am.

After the call ends, the music resumes. Now Cathy stops it, however, as she feels it is better to converse with the old fellow. He apologises for overhearing the conversation and butting in. Cathy does not mind, especially after getting useful help with her phone.

Besides, she loves talking to people. Beats music any time. Well, except for Weather Report, but now she can get back there when she wants.

The previous illustration seems common and perhaps even reasonable, but it also consists of a few interesting cases related to system behaviour. First, how is the incoming call presented in the first place? Typically, a ring tone is used to alert the user, in addition to haptic and visual output, but now the user is already using the audio channel and the device is placed in the bag so the vibration patterns and visual indicators are not noticed. Second, answering the call will pause the playback as they both use the same modality. Of course, the system could continue playback on the background of conversation, but that is not desirable as it hinders the conversation. Next, the user needs to access the calendar while being engaged in a call. Therefore, the interaction with the device should not affect the call. Furthermore, when the phone call ends, what is the state to which the system should return? Before the call, the music was being played, but on the other hand, the calendar application is open since the user interacted with it during the call. Design decisions such as these are task-dependent and must be made on a case-by-case basis.

6.6   The Effect of Device Category on UI

In Chapter 2 we classified mobile devices into three categories based on the number and expandability of their functionality. Evidently, these factors have an influence on the user interface too. In fact, the category of the device affects the UI greatly as the increased number of features require more flexibility from the UI (remember the feature race discussion at the beginning of this chapter?). For example, if the device is only dedicated to audio playback, the UI should reflect this aspect and provide fast and easy access to the core operations related to this task. On the other hand, the UI of a toolbox device should be flexible enough to support a wide range of different tasks. To make things even more interesting, all possible tasks for a toolbox device are not even known at the time of the device design (since the platform is open for further (third party) application development).

The user interface of dedicated media devices is normally minimal; only the controls for the most frequent actions involved in primary tasks are provided. This is emphasized by using hardware controls for direct access to key actions, such as a button for starting and pausing playback in an MP3 player, or a slider for zooming and a button for capturing in a digital still camera. The rest, secondary operations, are hidden inside some form of menu structure. The user performs only one task at a time, implying that there are separate states for capture and browse in a camera, or listening to a radio or MP3 files in an audio player. Hence, task switching need not be considered.

Dedicated devices do not have UI guidelines, as the proprietary platforms are not open for third-party application development. However, due to consistency and brand image, the device manufacturers extend their own UI conventions over a broad range of products. Also, often the devices of this category mimic the design solutions of other similar devices, which is obvious if comparing digital cameras from different manufactures, for instance.

Swiss army knife devices are more feature-loaded than dedicated devices, which implies that their UI consists of a large amount of multi-purpose UI components and physical interaction devices. The structure of the UI is dominantly hierarchy-based, but the increased number of technologies and features lead to a broader and wider menu structure for accessing functionality. In addition to generic interaction devices, the devices may have additional physical controls for stressing the key or differencing functionality. For example, some mobile phones have additional playback or capture controls. Support for multi-tasking is still limited, mostly due to the limitations of the computing platform (consider the use case above). Concerning third-party application development, the devices may have guidelines and conventions for user interface design, as they usually support technologies such as Java.

Nearly all UIs in the toolbox category devices blend menu selection-based interaction style with a WIMP GUI. These devices have many pre-installed applications and the user can (relatively) easily install new applications. Thus, a novice user feels that the UI is complex to master. On the other hand, the devices may offer an opportunity to organize and personalize the menu structure so that it is easy and fast to use. This also implies a difference between the toolbox category and the other categories: the UI structure of toolbox devices rely more on the concept of an application than the other categories (see Chapter 7 for more discussion on mobile applications). For instance, from the end-user's perspective, dedicated devices usually have "features" instead of "applications" and the same applies to most Swiss army knife devices. Toolbox devices are often preferred by experienced users who demand efficiency and effectiveness in a single package, specifically tailored to their purposes. Due to this and exploiting the concept of application, multi-tasking and task switching is almost as important as with desktop computers.

Toolbox devices usually provide platforms (such as Series 60, UIQ, or Linux) for third-party application developers. These devices typically provide comprehensive UI style guides, examples, and other documentation that aid in using the platform's UI components and conventions for UI design and implementation. In addition, toolbox devices are appealing targets for developers as they provide them with services, such as messaging and imaging. They also have more processing power and memory, better graphics and audio capabilities, and various connectivity features (such as, GSM, 3G, WLAN, and Bluetooth).

6.7   Summary

The user interface is a channel for the user to communicate their actions to the system and receive feedback when performing a task, that is, there is no other way to interact with the product. Therefore, in whatever way the UI fails in some aspect, it will have a negative impact on the product, which eventually reduces the use of the product and, ultimately, sales.

To create successful products, designers should know who the design is being created for, why it is done, and what is the best way to do it with the given technology? Realizing these goals requires a lot of work, including research on markets, trends, technologies, and competitors to check feasibility on several levels.

Getting the basics right is only the first step towards a successful UI, the next one is to define the product's users and tasks. There are many options available for this task, but most likely it will succeed with the aid of user-centred design methods . . . and with common sense. The user interface design involves dividing the tasks between the views, defining the structure and the UI components in a manner that follows all general UI design principles, platform guidelines and conventions, and is aesthetically pleasing to all the senses.

It is highly unlikely that any design is initially perfect. It must be validated and iterated with various usability engineering methods. Moreover, a triumphant UI design requires deep understanding of the technologies and how to use them in the best possible way.

Because mobile computing and mobile personal content are still in their infancy, there is plenty of room for introducing new and novel solutions. They can relate to the perceivable part of the UI (presentation techniques), to the system core (automation and adaptation metadata magic), or to interaction techniques. The user interfacing for mobile personal content experience calls for products that smoothly tackle one or more GEMS phases (together or separately) as a part of the device ecosystem.

6.8   References

Aaltonen A. (2007) Facilitating Personal Content Management in Smart Phones. PhD thesis for University of Groningen, The Netherlands, March 2007, 171 pp.

Aaltonen A., Huuskonen P., and Lehikoinen J. (2005) Context Awareness Perspectives for Mobile Personal Media. *Information Systems Management*, Volume 22, Number 4, Sept. 2005, pp. 43–55, Auerbach Publications.

Apple iPod. Available as http://www.apple.com/fi/ipod/ipod.html. Link checked 08.06.2006.

Aula A. (2005) *Studying user strategies and characteristics for developing web search interfaces*. Doctoral dissertation, University of Tampere, Dissertations in Interactive Technology, Number 3. Available online at http://acta.uta.fi/pdf/951-44-6488-5.pdf.

Bateman J. (1998) *Changes in relevance criteria: An information seeking study*. PhD dissertation, University of North Texas, School of Library and Information Sciences.

Baudish P. and Rosenholz R. (2003) Halo: a technique for visualizing off-screen objects. *Proceedings of Conference on Human Factors in Computing Systems* (CHI), pp. 481–488, Fort Lauderdale, Florida, USA, ACM Press.

Bederson B. and Hollan J. (1995) A Zooming Graphical Interface System Demonstrations. *Proceedings of the SIGCHI conference on Human factors in computing systems* (CHI) Volume 2, pp. 23–24, Denver, Colorado, United States. ACM Press.

Bental D., Cawsey A., McAndrew P., and Eddy B. (2000) What does the user want to know about Web resources? A user model for metadata. *Proceedings of the International Conference on Adaptive Hypermedia and Adaptive Web-Based Systems* (AH2000), pp. 268–271. Springer-Verlag.

Bertin J. (1981) *Graphics and Graphic Information Processing*. Walter de Gruyter & Co., Berlin.

Carrol J.M. (ed.) (2003) *HCI Models, Theories, and Frameworks: Toward a Multidisciplinary Science*. Morgan Kaufmann.

Cutrell E., Dumais S.T., and Teevan J. (2006) Searching to Eliminate Personal Information Management. *ACM Communications*, January 2006, Volume 49, Number 1, pp. 58–64. ACM Press, New York, USA.

Dix A., Finlay J., Abowd G., and Russel B. (2003) *Human-Computer Interaction* (3rd edn). Prentice Hall.

Drucker S.M., Wong C., Roseway A., Glenner S., and De Mar S. (2004) MediaBrowser: Reclaiming the Shoebox. In: *Proceedings of the working conference on Advanced Visual Interfaces* (AVI), Gallipoli Italy, ACM Press, pp. 433–436.

Fitts P.M. (1954) The Information Capacity of the Human Motor System in Controlling the Amplriture of Movement. *Journal of Experimental Psychology*, Volume 47, Number 6, pp. 381–391.

Hakala T., Lehikoinen J., and Aaltonen A. (2005a) Spatial interactive visualization on small screen. *Proceedings of 7th international conference on Human Computer Interaction with Mobile Devices and Services* (MobileHCI), pp. 137–144, Salzburg, Austria, ACM Press.

Hakala T., Lehikoinen J., Korhonen H., and Ahtinen A. (2005b) Mobile Photo Browsing with Pipelines & Spatial Cues. *Proceedings of INTERACT 2005*, LNCS 3585, pp. 227–239, Springer.

ISO, International Organization for Standardization (1998) Ergonomic requirements for office work with visual display terminals (VDTs), Part 11: Guidance on usability. ISO/IEC 9241-11: 1998 (E).

International Telecommunication Union (ITU), Telecommunications Standardization Sector. ITU Recommendation E.161 (02/01), Arrangement of Digits, Letters and Symbols on Telephones and Other Devices that can be Used for Gaining Access to a Telephone Network.

Jaaksi A. (1995) Object-Oriented Specification of User Interfaces. *Software – Practice and Experience*, Volume 25, Number 11, Nov. 1995, pp. 1203–1221, John Wiley & Sons.

Kamba T., Elson S., Harpold T., Stamper T., and Sukaviriya P. (1996) Using Small Screen Space More Efficiently. *Proceedings of ACM Conference on Human Factors and Computing Systems* (CHI), Vancouver, British Columbia, Canada, ACM Press, pp. 383–390.

Karpov E., Kiss I., Leppänen J., Olsen J., Oria D., Sivadas S., and Tian J. (2006) Short message dictation on Symbian series 60 mobile phones. *Proceedings of the 8th international conference on Multimodal interfaces*, pp. 126–127, Banff, Alberta, Canada, ACM Press.

Knoche H., McCarthy J.D., and Sasse M.A. (2005) Can Small Be Beautiful? Assessing Image Size Requirements for Mobile TV. *Proceedings of ACM Multimedia*, pp. 829–838, 6–12 November, Singapore, ACM Press.

Lehikoinen J., Salminen I., Aaltonen A., Huuskonen P., and Kaario J. (2006) Meta-searches in peer-to-peer networks *Personal and Ubiquitous Computing*, Volume 10, Number 6, Oct. 2006, pp. 357–367, Springer-Verlag, London.

Lehikoinen J. and Aaltonen A. (2003) Saving Space by Perspective Distortion when Browsing Images on a Small Screen. *Proceedings of OZCHI*, pp. 216–219, Brisbane, Australia, CHISIG, The Ergonomics Society of Australia Inc.

Lindholm C., Keinonen T., and Kiljander H. (2003) *Mobile Usability: How Nokia Changed the Face of the Mobile Phone*. McGraw-Hill.

Maltz D. and Ehrlich K. (1995) Pointing the way: Active collaborative filtering. *Proceedings of the 1995 Conference on Human Factors in Computing Systems (CHI)*, pp. 202–209.

Miller G.A. (1956) The Magical Number Seven, Plus or Minus Two: Some Limits on Our Capacity for Processing Information. *Psychological Review*, Volume 63, Number 2, pp. 81–97.

Motorola Ming – Technology – Business card reader. Available as http://www.motorola.com.hk/eng/motomobile/a/a1200/features03.asp. Link checked 08.06.2006.

Mullet K. and Sano D. (1995) *Designing Visual Interfaces: Communication Oriented Techniques*. Prentice Hall.

Myers B.A. (1996) User Interface Software Technology. *ACM Computing Surveys*, Volume 28, Number 1, Mar. 1996, pp. 189–191, ACM Press.

Nielsen J. (1994) *Usability Engineering*. Morgan Kaufmann, San Francisco.

Nokia 7710 Product pages. Available as http://www.nokia.com/nokia/0,65385,00.html. Link checked 16.01.2006.

Nokia Lifeblog. Available as http://www.nokia.com/lifeblog. Link checked 20.3.2006.

Nokia MiniMap Browser. Available as http://www.nokia.com/browser. Link checked 04.10.2006.

Norman D.A. (1990) *The Design of Everyday Things*. Bantam Doubleday Dell Publishing Group.

Norman K.L. (1991) *The Psychology of Menu Selection: Designing Cognitive Control at the Human/Computer Interface*. Ablex Publishing Corporation.

Palm. http://euro.palm.com/fi/en/products/index.html. Link checked 08.06.2006.

Palm b. Ways to Enter Data into a Palm® Device. http://www.palm.com/us/products/input/. Link checked 11.06.2006.

Perlin K. and Fox D. (1993) Pad: An Alternative Approach to the Computer Interface. *Proceedings of Computer Graphics* (SIGGRAPH), pp. 57–64, ACM Press.

Research in Motion (RIM) BlackBerry 8700c Demo. Available as javascript: popup ('/8700cdemo/container.jsp','8700cDemo','780','540','scrollbars=no','resizable=no'). Link checked 11.06.2006.

Robbins D.C., Cutrell E., Sarin R., and Horvitz E. (2004) ZoneZoom: map navigation for smart phones with recursive view segmentation. *Proceedings of the working conference on Advanced Visual Interfaces* (AVI), pp. 231–234, Gallipoli, Italy, ACM Press.

Rosenberg R. (1998) *Computing without Mice and Keyboards: Text and Graphic Input Devices for Mobile Computing*. Doctoral Dissertation, Department of Computer Science, University of London.

Sarkar M., Snibbe S., Tversky O., and Reiss S. (1993) Stretching the Rubbersheet: A Metaphor for Viewing Large Layouts on Small Screens. *Proceedings of the 6th annual ACM symposium on User interface software and technology* (UIST), pp. 81–91, Atlanta, Georgia, United States, ACM Press.

Series60.com (2005) Series 60 UI style guide (version 1.2). Available as http://series60.com/file?id=268. Link checked 07.12.2006.

Shneiderman, B. (1998) *Designing the User Interface* (2nd edn). Addison Wesley.

Siirtola H. (1999) Interaction with the Reorderable Matrix. *Proceedings of International Conference on IEEE Information Visualization*, pp. 272–277, IEEE Computer Society.

Sony Ericsson P990i Overview. Available as http://www.sonyericsson.com/ spg.jsp?cc=gb&lc=en&ver=4000&template=pp1_loader&php=php1_ 10336&zone=pp&lm=pp1&pid=10336. Link checked 11.06.2006.

Spink A., Bateman J., and Jansen B.J. (1999) Searching the Web: a survey of Excite users. *Internet research: Electronic Networking Applications and Policy*. Volume 9, Number 2, pp. 117–128.

Spink A., Wolfram D., Jansen B.J., and Saracevic T. (2001) Searching the Web: The public and their queries. *Journal of the American Society for Information Science and Technology*, Volume 52, Number 3, pp. 226–234.

Symbian (2005) Symbian OS: View Architecture v1.1. Available as http:// forum.nokia.com/info/   sw.nokia.com/id/3cab67ad-db01–400a-9467– 91b8be7ccbba/ Symbian_OS_View_Architecture_v1_1_en.pdf.html. Link checked 07.12.2006.

UIQ Technology (2002) UIQ Style Guide (version 2.1). Available as http:// www.uiq.com/files/UIQ_21_Style_Guide.pdf. Link checked 07.12.2006.

van Dam A. (1997) Post-WIMP User Interfaces. *Communications of the ACM*, Volume 40, Number 2, Feb. 1997, pp. 63–67.

Wilson T.D. (1981) On user studies and information needs. *Journal of documentation*, Volume 37, Number 1, pp. 3–15.

Chapter 7: Application Outlook

application device **game**
information **location** mobile people
player POI service sharing site user video

In the preceding two chapters, we provided an insight into the internal workings of reference software architecture for managing personal content, describing the essential technical features implied by identified user needs and requirements, and the limitations imposed by existing technology. Furthermore, we presented a similar approach to content management user interfaces. We showed the essential elements for designing usable interfaces for experiencing mobile personal content, while dealing with the inherent restricting features of small devices – limited input and output capabilities. Now we focus on the way the solutions are linked and how they are offered to users.

Still missing from our discussion, from a traditional software stack point of view, are applications. This is the software that exploits services provided by the platform, brings in the domain logic for dealing with related information, and applies user interface design principles to implement a user interface, thus making the functionality

Personal Content Experience: Managing Digital Life in the Mobile Age   J. Lehikoinen,
A. Aaltonen, P. Huuskonen and I. Salminen © 2007 John Wiley & Sons, Ltd

Figure 7-1. Creative MuVo TX audio player.

available.[1] Application development is a balance between desired functionality and usability – developing a feature-rich application with an intuitive user interface is a real challenge.

In the context of mobile personal content, the term *application* is slightly different from the traditional software-centric view. Dedicated devices, providing extensions to features and functionality, usually implement a small subset of the functionality in the GEMS model, which is targeted at one primary content type only. They also tightly combine both the platform and the application (and the device), to the extent that they may be inseparable (in this case, the term *appliance* is often used). A typical example is a small-screen MP3 player that is only designed to play audio content, such as Creative MuVo TX (Figure 7-1). In this case, the primary application is an audio player, the device being designed with audio playback capabilities in mind. However, it can record sound and, as a USB flash memory device, can also be used to store digital content. Typical of devices in this class is that the GEMS maintaining phase is usually performed elsewhere, such as on a desktop computer; the playlist is combined in the PC, and only then transferred to the player. Organizing the content in the player user interface is very limited.

---

[1] The term *application* often refers to a single block of executable code, which occasionally refer to external common libraries. However, an application may also be understood as a more loose combination of software modules that may be attached and detached dynamically, depending on the user task, to provide a combination of services suitable for current needs. An application or a service can also be implemented as a dynamic web page.

At the other end of the spectrum, there are toolbox-like multi-purpose devices that provide an open platform for further application development. Usually, devices in this category implement the majority of GEMS phases and can manage several different content types. Therefore, they may require a clear separation between the applications and the platform. A good example of these is current smart phones.

In this chapter, we consider both dedicated application+device designs, and multimedia computer-like toolbox devices with an open platform offering a variety of applications. Furthermore, we consider applications that are provided as web services. Though they are not mobile as such, in most cases they can be accessed by means of mobile devices.

What is common to all of these approaches is that the application, inclusive of the platform component or not, provides the logic and implementation of the personal content experience functionality, through a user interface.

Often the separation is not as clear as indicated above. Many vendors, such as Apple with its iTunes and iPod suites, provide bundled solutions as a combination of applications, services, and devices that work smoothly together, forming a coherent user experience. Most device manufacturers include PC software for the "M" phase in their sales package, such as photo organizers provided by camera manufacturers, and music organizers by media player manufacturers. Many have online services that add to the functionality provided by the primary application. A typical example is a camera manufacturer that provides an online album service for its customers, with the ability to share the album with friends and family, support for communities, and other functionality.[2] Hence, the GEMS phases that may be missing in the primary application can be performed through an additional media service.

This chapter adopts a case-driven approach. We examine the essential application design aspects derived from user needs and existing technology, and solidify them by presenting and analyzing numerous existing applications and services that highlight their particular aspects. While analyzing these cases, we not only tie them to GEMS in order to classify them, but also aim to identify the factors behind their successes or failures.

As discussed in Chapter 1, the domain of personal content is growing uncontrollably, implying that new devices, applications, and services

---

[2] See, for instance, Sony's ImageStation, http://www.imagestation.com/

are developed and offered to the consumers at a breakneck pace. There is a more or less functional solution available next to every imaginable GEMS action, to a wide variety of different content types. Obviously, most come with their own proprietary approach, incompatible with the others. Discussing a significant portion of these is impossible. The approach we choose is to point out solutions that we consider significant, either as of today or in the future, or that have gained some foothold in the marketplace, while hereby acknowledging the existence of the remaining 99.99%.

7.1 General Characteristics of Mobile Applications

As discussed in Chapter 3, many aspects of personal content management that have previously taken place in homes and offices are currently moving into the mobile domain, making it possible to create and enjoy content while on the move. Furthermore, mobility brings along new usage patterns and content types – and with them, necessarily, new application types. Many mobile applications are but crippled versions of their desktop counterparts, such as light versions of word processors or e-mail clients, but others are designed purely for mobile use.

As an example on how mobility changes the applications, consider PDAs. Many PDA models are not connected to any network, but are synchronized with the latest information once they are connected to their homebases (usually a PC). This non-continuous mode of updating shows up in many PDA applications, such as e-mail, calendars, news, etc., all based on updating the latest information only when connected. This also shows up in the application design, where the user is assumed to be aware of the various issues relating to synchronization. Besides PDAs, the same model applies, for instance, to mobile phones, laptops, and MP3 players.

Consumer software sales for desktop computers are an everyday business, with games and standard office applications usually ranking highest in top selling lists. On the contrary, the market for software for mobile devices has been far more limited, not least due to the many proprietary closed systems that do not support third-party applications. A direct consequence is that most users settle for the software provided with the mobile device and never bother to buy additional applications. With the increasing availability of open mobile platforms, such as Symbian[3] in many modern smart phones, this trend is slowly changing.

More than a decade ago, a significant mobile device, with an open platform for third-party application development, was Palm Pilot. Many applications were included out-of-the-box, mostly for Personal Information Management (PIM), such as a notebook, an address book, and a calendar, to get a good start with the device. However, the variety of third-party applications is astonishing. Online sites, such as PalmGear.com,[4] provide thousands of trialware, shareware, and freeware applications to download and purchase for Palm and Palm-based devices. At the time of writing, the most downloaded application in the top monthly chart was BigClock, a freeware extended clock application with timers and alarms, followed by a shopping list application. Also available for download are applications that make the device work as a TV remote controller, or that contain subway maps for hundreds of cities around the world.

Similarly to PalmGear.com, there are portals for other current mobile devices. A popular service for downloading software for Symbian-based smart phones is my-symbian.com,[5] providing application for Series 60 and Series 90 smart phones as well as for UIQ phones. At the time of writing, the most popular Series 60 download was Doom 7650, a Symbian port of the classic first-person shooter game, followed by infrared remote control (essentially similar to the software for Palm, discussed above). The third application in the most popular list is My Girls, which helps keep track of a woman's menstrual cycle, including when menstruation is due, and how fertile she is at any point in time. Mobile phones are indeed multi-purpose devices!

7.2 Location-Based Applications

A typical example of a mobile application is, not surprisingly, navigation. Since a mobile device goes with the user, *wayfinding* is an obvious and useful feature. Determining the user's location, or the location of the device, to be more precise, is referred to as *positioning*. A positioning service providing the user with information on their current location. Besides wayfinding, a myriad of different *location-based*

---

[3] http://www.symbian.com/
[4] http://www.palmgear.com/
[5] http://www.my-symbian.com

applications and services can be designed and deployed, including tagging, information services, and games.

In this section, we will look at location-based applications from the personal content point of view. This approach provides us with two options: either the location information is considered personal content as such – that is, all GEMS actions can be potentially applied to location information directly – or, alternatively, location information is considered a part of the context (that is, metadata), be it the user context or the file context, and is primarily used to enhance other types of personal content. Nevertheless, location information is in many cases a major building block of mobile personal content experience.

Put simply, a location-based application knows its current location, and can adapt to it. A location-based application can also be referred to as a *location-aware application*.

The most common positioning service to date is the satellite-based Global Positioning System (GPS) (Getting 1993). The position can be acquired with a GPS receiver. The accuracy of a GPS service is usually within a few metres, but with differential correction it can be brought down to a few dozen centimetres. GPS is governed by the US military; Russia also has their equivalent system in place, GLONASS, while the EU is planning to launch a similar satellite-based positioning system, called Galileo.

In addition to GPS, there are other positioning services available. As an example, the current cellular network is capable of determining the location of a connected mobile phone. There are various techniques to obtain the location coordinates but, regardless of the technique used, the positioning accuracy is much lower than with GPS. The accuracy depends on the technique used, as well as on the cell size, and is typically hundreds of metres. Similarly, wireless local area networks (WLANs) can be used to triangulate the user's current location; if the location of the WLAN access points is known,[6] the location of the device can also be determined.

7.2.1 Point of Interest

A central concept and an explicit content type related to location-aware applications is a *Point of Interest* (POI). A POI is defined as an "important, distinctive environmental feature". The term *landmark* is often used interchangeably. However, in the context of personal content, the definition of a POI can be broader – a POI can be a loca-

---

[6] Which is rare at the moment. Building a WLAN-based navigation system requires initial measurement of the location.

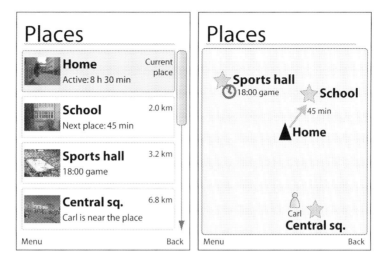

Figure 7-2. Some user-defined POIs in two alternative views.

tion with a specific importance and personal meaning to the user (for instance, home or school), even though it may not be distinctive as such.

In essence, a POI is a piece of information that contains the location reference, be it GPS coordinates, a cell ID, or any other specifier, and some information related to their location. The simplest form of a POI is the location reference and a name.

The POIs can be shared, combined with those from others, and so forth – thus a POI is in no way different from any other kind of personal content. As a consequence, there may also be POIs with different privacy levels, some being private, while others are shared with a restricted group of people. Some POIs can be in the public domain, being created and perhaps sold or traded by third parties, such as communities.

The concept of a POI can be extended to moving objects, which results in POIs that could be vehicles, pets, or other people. With these dynamic POIs, an absolute location is replaced with a logical or semantic location such as in the case of vehicles ("I'm in a train").

A POI is often explicitly created content, even though automatically created POIs also exist, such as waypoints along a suggested route.

7.2.2 Wayfinding

Navigation, or wayfinding, is the process of finding one's way from the point of origin to the destination. Usually, this can be understood as a

Figure 7-3. Magellan eXplorist 600 (Photo: courtesy of Tero Hakala).

series of actions that are to be carried out at specific locations; these action-location combinations can be referred to as wayfinding decisions. Once all the wayfinding decisions have been successfully performed, the destination has been reached.

Traditionally, navigation has been carried out with a map and a compass. Using a map is not a trivial task, since cognitive processes such as relating the scaled 2D map to the 3D real world, and finding the correct orientation, are required. Essentially, map use can be defined as "the ability to relate a two-dimensional map to the three-dimensional world" (Blades and Spencer 1987), that is, an ability to match the terrain model and map model in one's mind. By applying the latest technology, the process of navigation can be greatly facilitated, to the point that no map is needed, only the required wayfinding decisions being presented.[7] For a thorough discussion on how maps work, see MacEachren (1995).

A handheld GPS navigator, such as Magellan eXplorist 600, is a dedicated device targeted at wayfinding and route planning (Figure 7-3). The device is equi pped with an integrated GPS receiver and a screen, which is capable of displaying a map of the current region, augmented with some additional information, such as a recommended

---

[7] Navigating devices are increasingly common in vehicles; however, we do not consider vehicle navigation as a separate topic, but focus on personal content regardless of whether it is related to pedestrian or vehicle navigation.

route. The map information is usually stored on a memory card, which allows purchasing new maps and, by switching to another card when necessary, using the device anywhere. Some handheld GPS devices may also include secondary features, such as Garmin GPSMAP 60c, which contains information for geocaching, GPS–based treasure hunting,[8] and a few geolocation games. A barometer and a compass are other typical hardware additions to a GPS receiver.

For other devices, such as mobile phones and palmtop computers, separate GPS modules are available. For instance, Nokia LD-3W is an external GPS module that connects to the phone wirelessly via Bluetooth. The location information provided by the GPS receiver is then available to application developers. The decrease in price combined with increase in functionality has boosted the sales of personal handheld navigators, making them common in outdoor activities such as fishing, trekking, and hunting.

As far as wayfinding is concerned, there are several kinds of personal content created and used, either explicitly or implicitly. The targets (final destinations) are typical examples of user-created content, whereas the routes are often suggested by the application, even though in many cases the user may define the preferences for choosing a particular route. Similarly, waypoints (sub-targets along the route) are implicitly created. Furthermore, POIs are closely related to navigation as, for instance, user-created waypoints.

Typically, personal content related to wayfinding is not heavily maintained. The content is created either explicitly or implicitly. It is then enjoyed, in many cases implicitly, such as by reaching a target, and only occasionally shared. The expected lifespan of wayfinding-related personal content is often short – when moving in familiar environment, navigation is rarely needed, and once a destination has been reached, there is usually no need to store the related information, unless there is a need to navigate to the same destination in the future. In many cases, the importance of the stored information is low, since the same information can be easily re-obtained.

7.2.3 Annotations

Annotations, or tags (or virtual graffitis), are a typical form of location-based personal content. In its simplest form, an application may allow adding location-dependent notes that can be later locally retrieved and presented by overlaying them on a map. More advanced annotation concepts provide tools for sharing and publishing the annotations, that

---

[8] http://www.geocaching.com/

is, attaching the note to a location so that other users can see it while in the vicinity. An example of an annotation application with global coverage is Google Earth,[9] which among other things allows creating annotations on any location. The basic version of Google Earth is not classified as a location-based application, since it is not aware of its own current location. However, there are more advanced versions available, such as Google Earth Plus, which adds support for GPS devices.

There is a slight difference between a POI and an annotation. Put briefly, the purpose of a POI is to name a location, whereas an annotation describes the location. Furthermore, the annotation is in many cases not tied only to the location, but to other context information as well, such as time. In essence, there is usually a *contains* relation between a POI and an annotation.

By applying simple annotations, applications, such as fishing journals that are able to record the location of a bite, can be developed. Many stand-alone GPS receivers allow such limited annotations as a simple extension to mere wayfinding and route planning.

A textual annotation attached to a location is a good start, and can be considered as personal content. However, as discussed in Chapter 3, the relations between content items are an essential element that brings the separate pieces together and gives them meaning. Therefore, it would be beneficial to be able to store, along with the annotation, a picture taken of the proud fisherman, or the text message sent to the friends who were not able to attend the trip. Not to speak of combining the content related to the situation from all the participants. Most stand-alone devices do not allow such integration of information, nor connecting to external devices and synchronizing it.

7.2.4 Location as Metadata

Not all location information can be seen as independent content. Often, location information is a part of the context – the current user context, or the file context at the time the file was created. Increasingly, many content-related applications allow storing location as a metadata item.[10]

---

[9] http://earth.google.com/

[10] For the sake of academic discussion, one could argue that all location information is metadata – an annotation, for instance, is a textual or pictorial piece of content, and the creation location is its metadata. However, we wish to make this distinction, since a location-based annotation loses its meaning if the location information is omitted. The line between an "annotation" and "a piece of content with location metadata" is indistinct at best.

Consider a camera phone equipped with a GPS receiver or an equivalent positioning technique. An obvious application is a photo album, which stores the images and includes location information in each photo. This is exactly what Nokia Album does.[11] The photos can be grouped based on location; furthermore, location can be used as a search term. Many studies exist on augmenting photos with location information (Davis et al. 2004; Toyama et al. 2003; Sarvas et al. 2004; Naaman et al. 2004).

Obviously, photos are not the only content type that benefits from the inclusion of location metadata. In fact, any piece of content would benefit if location information was included. Consider, for instance, editing a text document. Each time you modify it, the file system automatically stores a timestamp along with the file, so you are able to see when the file was last modified. Why not include location information in a similar manner? It would then be an inherent characteristic of any piece of content, and as accessible as the timestamp is today. In essence, such information would be extremely useful in searching, even though the content itself might have no relation to the location that was created or last modified.

7.2.5 Location and Communities

The ability to access another user's current location, either continuously or by defined intervals, is referred to as *tracking*. Tracking allows many social applications, such as group communication and planning, and location-aware presence – that is, the ability to check a friend's current social status, including availability and location (Lehikoinen and Kaikkonen 2006).

In essence, social location information can be divided into two categories:

1. Where's everybody?

2. Is anybody there?

Tracking the location of other people is a strong candidate for making location-aware applications a wild success. Combining the features discussed in the preceding sections, such as the creation and sharing of POIs, one can effortlessly envisage concepts that allow creating personalized maps, sharing them with friends, and, by including tracking the location of the group members (dynamic POIs), *ad hoc* planning. The options are indeed numerous.

[11] http://europe.nokia.com/nokia/0,6771,58681,00.html

AARGHH! MY HAT!

Phew, that was close. The police had already resorted to tear gas to disperse the riot, and rubber bullets were probably next. Bob had managed to duck and not get hit, but a painful stick blow landed on his back when the police came roaring through the street. Two of the armoured officers were running around with tasers. That was bad stuff. Bob was just barely able to escape by jumping over the fence to a parking lot. He landed hard and painfully, dropped his helmet on the concrete, one microphone broken, but the recording seemed to be still okay, so he just popped it back on his head.

Not a good time now to review the footage. Bob needed to get an overview of the situation. Where's everybody? He sent a tracking call to query the locations of his cell friends. Within a minute he had a map. Most everybody seemed be in the front line where the bus was burning, some people were nearby, Jessie on this very street and probably going to get the gas soon. Five members of the cell were missing, maybe indoors without data coverage. Hope so. Their last locations were near the warehouse that served as their makeshift headquarters. Bob decided to head that way to get replacement batteries for his helmet. A lot of juice would be needed to get this crazy night out to the world.

In addition to tracking, there are alternative ways to support the social aspects of location information. For instance, cellspotting[12] is a web site that allows the users to find information attached to their current location by other users, and also add their own content (annotations) for others to see, by using a client application installed in their mobile terminals. Should you be unable to find any previously stored information of your current location, by adding your own content you become a "discoverer" of the location. As the name suggests, cellspotting uses location information obtained from the cellular network, also allowing tracking friends, and exchanging information with them.

```
COMMENT: ABOUT CELLSPOTTING

For many people, location information means navigation.
It probably answers the security need of classic Maslow's
hierarchy, as many fear getting lost. For me the more
interesting location questions are what I'm doing here
and what's around me? This is what makes cellspotting
quite addictive. It's fun to check whether someone has
managed to spot the location or do I get the honour of
naming the location. Compared to other popular location
"games" like geocaching, cellspotting doesn't require me
```

---

[12] http://www.cellspotting.com/

to go to any specific spot, but without travelling there's obviously nothing to spot. In this way the spotting becomes a virtual world that enhances travelling. There are some social features included, but without active spotters the service would be completely useless. This notion of supporting a growing community and sharing content makes it the final twist to the spotting. My finest moment as a spotter was when one of the authors of this book came back from a trip to Lapland, saying: "I tried to spot there, but someone named Jylppy had already spotted all cells." I don't own the cells in any way, but I have left my marks for others to spot.

Juha Kaario
Senior Research Manager
Nokia

(Nickname "Jylppy" in cellspotting, number 10 in top ten spotters list in autumn 2006)

There is an obvious inherent risk related to tracking – privacy. Allowing anyone to query one's current location is not considered a viable option, requiring that some kind of control mechanisms and access rights schemes are incorporated. Defining who can access your current location, with what accuracy, where, and at what time of the day, either results in huge challenges is developing a simple enough user interface, or alternatively a system that effectively prevents its use altogether. Furthermore, by collecting and combining location information – or their absence, for that matter – easily results in a "robber's heaven" (the ability to check whether the inhabitants of the target are at home, and where the neighbours are). There are several side-effects to allowing location tracking, all of which undoubtedly are not discovered until such a system is used by millions. After all, the users – that is, all of us – will always find the best ways to abuse any system.

SHE'S NOT THERE

Eddie. Eddie wants this, Eddie wants that. Eddie decides this and then they do it. Angie just follows. Why is it that somehow it's always him behind the wheel? Angie just thinks it's maybe easier to let him have his way. He's such a difficult person sometimes. What wouldn't a girl do for her guy?

Now, Angie had once again received a place ping from Eddie. She would normally have okayed it, as she's got nothing to hide from her dear boyfriend. This time though, she hesitated revealing her location, which was by

> the school cafeteria. Eddie probably wanted to get together for the break, but today she would prefer to lunch with Liz. Besides, it might be a good idea to let him feel a touch of doubt. This guy is too used to having his way.
>
> So it's Angie in command now. Here, click the "Decline" button. Sorry Eddie, your little cutie beauty blondie baby's not there for you, not this time. Now, where's Liz . . . ? Contacts – Find – Gals – Lizzie – Locate . . .

7.2.6 Other Applications

For multipurpose devices, location information can enable numerous different applications, including location-aware browsing and searching, games, location-based reminders, location-aware presence information, and so on. Most of these applications deal with personal content more or less directly.

Location-based prediction is an interesting feature – by learning the user's daily routes, the application can be used to predict where and when the user will head to from the current location (Ashbrook and Starner 2002). The information used in prediction is one kind of implicitly applied personal content: the locations you usually visit throughout a day, and the locations you probably go on to from there.

7.2.7 Discussion

Positioning is destined to become a standard feature in mobile phones. Gartner (2006) estimates that by 2010, almost 40% of all handsets will support GPS. That amounts to more than a billion devices. This market volume will drive component costs down, which in turn allows the introduction of GPS functionality in many other kinds of devices. All MP3 players and video viewers may ultimately feature positioning, although it is not yet clear what the benefits of that function will be. For example, game consoles could use GPS to flock nearby people together in virtual worlds. Is that a desirable feature? Maybe not – but some other positioning feature will become the killer app in a game console. Time will tell.

Generally speaking, most location-based applications and devices do not allow much content management, besides some rare actions in the "G" section, such as creating annotations. Especially enjoying location-based personal content, except for games, is rarely seen. The same holds for sharing, which is an obvious extension to enjoying. One could effortlessly envisage services and applications that tackle this aspect of location information. Approaches that are marginal as of today, such as cellspotting, may well show the way towards this in the near future.

The importance of sharing should not be underestimated. Even though many location-aware devices of today are not connected while on the move, primarily focusing on wayfinding, this is likely to change in the near future, especially once GPS-enabled smart phones, such as the Nokia N95,[13] start to find their way into stores. Once the connectivity is there to support the social aspect of location information, many new avenues for applications and services will open up. What the application developers should be aware of, especially where connected applications is concerned, is the privacy issue as discussed above.

THERE'S ONE IN EVERY CROWD

Dave and the boys are in the local pub. As usual, a hard day's night. That day they had re-installed the pipes for the mayor's office, which was a dirty job, as the house was Victorian or older. Nothing that a pint or two of bitter wouldn't cure, though.

The new guy, Ritch, had got a new cellphone and was showing it off. He explained that as they had their first-born baby, he wanted to record as much as possible of those early years. So he had got himself a phone with video functions. There he was now in the pub, showing videos of diaper changing. Well, Dave would prefer the darts and the pint, given the choice. Besides, if he had a videophone, what would he film?

Now Ritch was explaining something about those location functions. Not for Dave, that, either. Ritch showed how he could query the server to see it there were any interesting people nearby. Interesting people? Probably newbie fathers who wanted to talk diapers, Dave mused.

Ritch asked Dave what kind of people he would like to meet. "John Lennon", Dave quipped, "but he's probably not on Phoney". Ritch was not taken aback, and insisted that Dave would try the location functions anyway. Dave finally agreed, and asked Ritch to query the nearby people to see if any of them were into betting. Is anybody there?

To Dave's surprise, dozens of id's showed up instantly in Ritchie's phone. Dave tried looking at those strangers' mobile pages. For instance, there was this banker in the Dog and Crown down the street who used the nickname "Thorough-brewed". That must have been a nod to his main hobbies – pubs and horses. And there was even another B'ham lad who claimed be into races. That guy was on his way home but mentioned a nearby pub as his fave place. Hmmm.

Dave preferred talking to people instead of cellphones, but maybe there was something to this location thing. Anyway, that would have to wait, as now it was darts time. Ozzie! Another round for the boys, please.

---

[13] http://www.nokia.com/A4136001?newsid=1077775

Currently, in many cases, excluding POIs and games, it is only through indirect use of metadata that location information really starts to enhance personal content experience. As the positioning capability will become a more standard feature in many mobile devices in the near future, it is advisable to start providing content-intensive applications with support for location-related metadata and management, such as organization, categorization, and searching. Considering the location stamp a standard file property, not unlike *file creation* and *last access* timestamps are today, is the first step to take.

7.3   Sharing and Communities

As pointed out in Chapter 3, sharing is tightly related to personal content experience. The sharing of information can create a community, that is, people who form a group by sharing the same subject of interest. The relationship between personal content and communities then deserves a closer look. In this text, we focus on virtual communities.

There are communities of different sizes. Some only have a few members, while others have millions, some being more active than others. Some are closed – either they do not allow new members at all, or the new member candidates are carefully chosen – and some are more open. Nevertheless, it is about people getting together for a specific purpose.

7.3.1   Content Sharing

Content sharing sites, such as Flickr[14] for photo sharing, create so-called implicit communities. Anyone can view images in Flickr, but to post images there people have to register – to become part of the sharing community. "I'm in Flickr," we hear people say, much as if they would say "I live in Nottingham." Inside this loose community, more tightly knit families are implicitly created between people who comment on each other's photos. The community is only implicit, though, and the sense of belonging is not a major factor among Flickr users.

From the viewpoint of personal content, Flickr can be simply considered an online photo album with convenient uploading and hosting of images, with added functions for presenting, linking, sharing, commenting, and tagging. The interaction on the page has been streamlined for these key functions. Single clicks suffice to carry out most

---

[14] http://www.flickr.com/

Figure 7-4. Posting an image from a mobile phone to Flickr with LifeBlog (reproduced by permission of © Nokia).

tasks. Content can be conveniently uploaded to the site from mobile devices. Figure 7-4 shows how an image is posted to Flickr, with an added title.

Tags are simple text strings associated with the images, which creates a form of metadata. People can add arbitrary tags freely. Millions of tags have been defined so far, with the most popular tags[15] involving tourism, scenery, pets, families, parties, weddings, holidays, and so forth.

Tags create free-form collections of metadata, known as *folksonomies*, as opposed to taxonomies in the knowledge representation community (section 4.3). This grass-roots approach to metadata appears attractive, as tags are simple to contribute, simple to understand (for humans), easy to index and search, and readily embedded among HTML. However, since the metadata symbols are solely text, their semantic meaning is only clear to humans. For instance, a city might be tagged with "New York", "New", "York", "NY", "New York (NY)", or combinations thereof. These equivalent tags appear as separate ones to Flickr. This is a problem for automated analysis, but many of the users of the site will not care, happy to find their images with a process they are used to – keyword search.

7.3.2 Content Rating

An interesting loose community has developed around the *Hot or Not!* Phenomenon.[16] The idea is very simple: you submit your photo (your portrait, to be exact) for other web site visitors to rate, or you rate

---

[15] The most popular tags are listed in: http://www.flickr.com/photos/tags/
[16] http://www.hotornot.com/

photos sent by others. Each photo is given a hotness rate on a 1–10 scale. Furthermore, a participant may check the score they have earned; while doing this, the timestamp is recorded. This timestamp is also visible for those viewing the photos. This way, you not only rate a photo and can see how others have rated it, but are also able to check when the person in the photo has checked their score. Contacting the person in the photo is also possible, as in dating services. There is also a scoreboard that allows a group of people to compare their scores. At the time of writing, more than 23 million photos have been submitted, and more than 12 billion votes cast. There are numerous other rating-related web sites, such as those for rating cars, pets, gardens, or even teachers.[17]

What can we learn from such a community? Perhaps the ingredients for a global concept of hotness? Or perhaps the site reflects the human desire for 15 minutes of fame? Or our inherent will to rate everything we experience? Or maybe it is just a game without any underlying truths? Probably all of this, and more. Nevertheless, rating sites includes many factors important in personal content experience, even though this may not be immediately evident. Rating alone creates a link between the one who gives the rating and the content. Especially relevant is the site from the photo submitter's point of view, as the content can be easily shared. Metadata will be included, such as the ratings and photos provide links to people; and so forth.

Even though the concept of rating photos is not in any way mobile, envisioning services that provide the same functionality while being mobile is straightforward. Why not rate objects while on the move, using the built-in camera of an imaging phone and sending the photo instantly to the rating service for others to review? In fact, such services do exist today. One of them, of a somewhat dubious nature, is Mobile-asses[18] that specializes in rating photos with a restricted, very specific kind of content.

7.3.3 Self-Expression

MySpace[19] was, at the time of writing, the most prominent social networking site among youth (particularly western 15–25-year olds), with

---

[17] For a comprehensive list of rating sites, see http://www.ratingspot.com/

[18] http://www.mobileasses.com/

[19] http://www.myspace.com/

daily page view counts reportedly surpassing even Google. Every registered user of the site has a *profile* – a personalized page with multiple elements and functions, including personal data, guestbooks, messaging, links to other users' pages, rating, pictures, videos, music, animations, and so forth. The pages are highly customizable and typically rich in multimedia elements.

The key functions of the site are adding links to other user's pages, commenting on their pages, and customizing personal pages. The social networks on the site evolve through these functions, and can become dynamic with additions of new people and deletions of others. Besides individual users, a multitude of non-personal pages have been created for musicians, bands, products, even for software packages.

The primary motivation for MySpace use seems to be participation in juvenile "hanging out" with peers (Boyd 2006). The site also serves as a channel for self-expression, as people often customize their pages with extensive numbers of personal elements, such as pictures, memories, confessions, and even their biographies. Moreover, a typical profile displays lists of personal preferences, with the likes of "My fave food," "The worst place ever", "Best music", and so on. Much of the discussion on the site consists of commenting on the contents of the profiles.

Is MySpace personal content? Undoubtedly, yes. A MySpace page is the projection of self into a public image. This image is then used as a tool for interaction with other people. The image in this sense consists of publicly visible items that are of particular importance among youth – appearance, preferences, media, links, and friends. Even though some the contents of the page are borrowed from elsewhere, they are also personal in the sense that they (supposedly) talk about the owner of that page.

7.3.4 YouTube

The YouTube[20] web site allows people to share their own videos. Registered visitors are able to upload, share, and tag their own clips, and view, rate, and comment on the clips of others. Users sharing a common

---

[20] http://www.youtube.com/

interest can create groups, subscribe to channels, and so forth. At the time of writing, there are around 100 million views on YouTube daily, with around 65 000 new video clips uploaded per day. YouTube has nearly 20 million unique users per month.[21] In addition to the web site itself, YouTube videos can be embedded into external web sites, blogs, or MySpace accounts.

YouTube is an extreme example of sharing personal content. In effect, with YouTube the term "broadcast" could also be used, since by default the shared videos are available to the general public.

Sharing the material on the Internet can create a virtuous circle. The growth of YouTube demonstrates this effect nicely. As a video clip is seen by more and more people, these people in turn contribute comments and new material. The material resides in the Net, so people can send the links to the videos from almost any place, including mobile devices. As a consequence, while previously people would show their home videos to friends, they now pick out the best bits and upload them to YouTube. The site can accommodate video material from all parts of the Long Tail – from home videos seen by just a few people, to quickly spreading pop scandals of global reach.

Looking at the list of most popular videos on YouTube reveals a range of content types. Funny clips abound, much like they do in candid camera and "silly world" TV show formats. Music videos are popular, despite the copyright risks. Great TV commercials meet new global audiences. Amateur films and art acts are much uploaded, if not downloaded. A particular class of user-created content are "jackass" type stunts, where ordinary people attempt funny, silly, or just plain dangerous tricks, such as driving a bicycle down a hill backwards. This illustrates our need to get that 15 megabytes of fame.

7.3.5 Video Sharing Comes of Age

Video sharing is now one of the most rapidly spreading Internet applications. People are uploading videos by thousands every hour, and downloading them by millions. Usually this is carried out with PCs, but mobile devices are now catching on.

Illegal sharing of commercial video material has been the norm in P2P networks for some time. Mobile TV broadcasts are starting. By the end of 2006, Google had famously acquired YouTube, for 1.65 billion USD. All in all, we can say that video has become a viable data type in the Internet – if not yet on the mobile Internet.

---

[21] http://www.youtube.com/t/fact_sheet

Previously, video material was usually too large to be carried over typical modem connections. There were exceptions, such as stamp-size news broadcasts, or film trailers that were cleverly compressed and hand-optimized. Since then, network bandwidths have increased by a few orders of magnitude. It is sometimes said that DSL speeds of 1 Mbit/s or more allow the average person to concentrate on the video content and not on the limitations of the network. Such speeds, and better, are now available to major populations of PC users globally. While not yet the norm in mobile networks today, HSUPA and similar standards will bring equivalent speeds to handsets.

Video coding has also improved.[22] Earlier standards, such as MPEG-1 and MPEG-2, have since been surpassed in efficiency by DivX (and Xvid), H.264, and others. The improvements have made it possible to fit a full-length movie of two hours or so into 1 GB of memory, with quality that the average consumer considers satisfactory. This amounts to roughly 1 MBit/s (with audio), which can already be streamed in real time over the Internet, or allocated to downloading a film in a few hours.

The first sites to pioneer large-scale video distribution were dealing with adult material. For those audiences, poor viewing quality was often acceptable. P2P file sharing sites were first flooded with MP3 music and pirated software, but today video material is abundant. The most popular downloads in Pirate Bay, an illegal file sharing site, include mostly video material: films, TV shows, adult content, and a wealth of recent software and music releases.

For a long time, it has been possible to digitize video content with the various video capture devices available for PCs. However, the users of those devices have had to put extra effort into converting their analogue footage to digital files. This extra step has hindered larger audiences from sharing their videos. In contrast, however, many current devices store their video content in digital formats that can be readily manipulated. Digital TV receivers, hard disk TV recorders, digital cameras, solid state video cameras, and mobile phones all produce digital files that are ready to be sent (albeit sometimes with extra compression) over networks. This immediacy is, we feel, a major factor behind the surge of video content.

---

[22] In fact, so much better and tighter, that audio seems to consume more than half of some movies circulating in P2P networks. It seems that many people are prepared to cut down on video quality to keep the size down, but prefer to keep the surround sound experience (almost) intact. This will (again) change, as new surround sound compression methods are now emerging.

What makes YouTube so special? At least the site offers a better user experience compared to the average commercial Internet site. It supports various browsing modes – search by keyword, by popularity, by related content, or just plain luck. Comments, groups, and topics can be created freely. There are no registrations to worry about, if you just want to search and download. If you want to upload material, you do need to register, but the service takes care of most other trivia, freeing the user from potentially tricky tasks involving codecs and bitrates.

Perhaps the key success factor, though, is the unrestricted nature. YouTube does not force its users to behave in a certain way, but rather applies comparatively loose suggestions. It does not tell the users what to do. This makes people hang around and experiment with the site.

In a way, a YouTube upload gives added semantics – that is, meta-data – to the original clip. Currently the metadata is not too structured, residing mostly in the form of tags, comments, and contacts. Automated metadata harvesting could, and undoubtedly will, be added to the site without major changes to the user interface.

Are YouTube videos personal content? In our view, yes. The source material may not be, but that act of posting a video to YouTube is personal content. It creates a point of reference from which other people can provide commentary, further material, and sharing the post. The video post becomes attributable to the uploading person. Even though the uploader's identity may be imminent for a while, it can quickly get lost, as copies of the material are created and then reposted. Still the original post remains, and the uploader can later refer to it as "my post".

However, YouTube videos are usually not private content. Not many people would upload their sensitive personal material to the Net. Different sites and channels are needed for sharing content for more tightly controlled access. Then again . . . given that there are zillions of clips available, maybe privacy risks are a lesser concern, after all. Witness the following conversation from the readers' pages of the *T3* magazine (December 2006, p. 45):

Q: I fancy recording a birthday video greeting for my mate, and thought about putting it on YouTube, but don't want everyone to have a go at it. Is there a better way?

– [A *T3* reader], Swindon

> **A:** When uploading a video to YouTube, you've got the option to make it private and only seen by people you invite, but it's a bit awkward sending out and having these invitations accepted. It's so much easier just e-mailing a link which people can click. So we'd say go public with the video anyway just before the event, then take it down afterwards if you're worried or embarrassed. Random viewers will almost certainly never find it amongst the constant deluge of new uploads. In fact, many well-wishers rely on this fact and make their private greetings public on YouTube for ease of playback. Try searching for "Happy Birthday" and your mate's name on YouTube and you may find a hilarious alternative that you could send to them!

There are also third-party tools developed specifically for use with YouTube. For instance, SceneMaker, a video-sharing tool from Gotuit, identifies scenes within videos from YouTube and Metacafe, and allows sharing the scenes only.[23] The scenes can also be tagged to better identify the content within the videos.

Schwartz (2006) assumes that the YouTube buyout by Google was first and foremost due to the market size and the fact that who has the most viewers in video sites wins the game at the end of the day. He also points out that 99% of the content in YouTube does not even have the slightest quality. And that, after all, is not at all important, as long as advertisements can be sold. Another viewpoint could be that the social networks are worth it.

7.4   Games

Today, playing (computer) games is immensely popular,[24] and as a business domain games are expected to grow twice as fast as the movie industry, reaching half of its volume by 2010.[25] Therefore, even though games are not often included when content is discussed, we see it as one of the most important domains of GEMSing personal content. In this section, we will present the most important content types related to games, and discuss the emerging game types – mobile and pervasive games.

As pointed out in section 7.2, location-aware games are a typical form of recently emerged new application types. Even though games themselves are not personal content as such, the content created for

---

[23] http://blog.wired.com/monkeybites/2006/12/slice_tag_and_s.html
[24] A study finds that among American teens, three out of four play "videogames" (HGVNU 2006).
[25] http://www.businessweek.com/innovate/content/jun2006/id20060623_163211.htm

games and during playing a game in many cases are. Furthermore, an interesting notion related to gaming is "abusing" the game technology for totally other purposes, such as creating personal content that is not related to a game at all. Next, we will briefly address these topics.

Computer games can be classified based on several criteria. Typically, they can be divided into genres – action games, strategy games, and so forth. Furthermore, some games are designed for a single player, whereas others are multi-player games. Games may be mobile or targeted at stationary use.

When discussing games, a phenomenon not to be bypassed is Massive Multiplayer Online Games (MMOGs).[26] The most significant subgenre is Massive Multiplayer Online Role Playing Games (MMORPGs). Millions of players have subscribed to games such as *World of Warcraft* (by far the most successful online game ever in the western world), *Ultima Online*, and *EverQuest*. In Taiwan, China, and Korea, MMOGs are even more popular than in the western world – games such as *The Legend of Mir 3* may have close to one million subscribers online *simultaneously*. *The Legend of Mir 3* is considered the largest MMORPG in the world. As a result, rich communities, guilds, and tribes have been born, accompanied by huge amounts of personal content in a myriad of different forms, including guild histories and structures, communication histories, and the like.

7.4.1 Mobile Games

Of special interest in the context of this book are mobile games. Non-electronic mobile games are not a new phenomenon. Since ancient times, people have taken games with them, and played wherever. Consider an obvious example of a traditional mobile game – a deck of cards. It is extremely mobile (pocketable), very versatile, and both single- and multiplayer games can be played. Even though this analogy may seem trivial, there is yet another factor that reveals an essential factor of mobile games, digital or otherwise: the *ad hoc* nature of gameplay.

The first electronic handheld gaming devices were equipped with LED displays, among them the Mattel Auto Race (1976). The player interacted with the game by pressing buttons to move their car (a bright dot) and to avoid obstacles (less bright dots). During the 1980s, LCD displays replaced LED displays in handheld games. However, the games were still extremely simple: the user controlled a digital char-

---

[26] For a thorough analysis, see http://www.mmogchart.com/

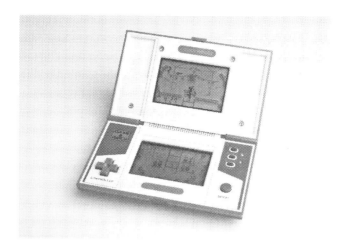

Figure 7-5. A Nintendo Game & Watch series game, GreenHouse (Photo: courtesy of Tero Hakala).

acter that was capable of a restricted set of movements in a static non-electronic background. One device contained one game only, with no variations. The most well-known handheld games of the LCD display era are the Nintendo Game & Watch series, such as *Donkey Kong* and *Greenhouse* (Figure 7-5).

Later, more advanced handheld game consoles emerged, among them Nintendo DS, Game Boy Advance, and Playstation Portable. These devices were not interesting as far as personal content is concerned – the only piece of non-static information being the score. And in most cases, the number of tries left prior to game over (i.e., lives).

Since those days, mobile games have diverted into two mainstream directions: handheld gaming consoles, and games in mobile phones and other devices, which are not primarily targeted at gameplay. From a personal content point of view, smart phones are more interesting, since they allow many GEMS actions.

Unarguably the best-known game in a mobile phone is *Snake*, introduced first in the Nokia 5110 in 1997 (Figure 7-6). It is a typical example of a mobile game that has no specific mobile elements.

Most mobile games of the *Snake* era are low-tech and usually played for a few minutes at a time when on the move. Such games are also referred to as *casual games*, and are considered the most important form of mobile games at the time of writing (Hyman 2005). Casual games can be challenging, but easy to access and quick to learn.

Mobile games that exploit the features of a mobile phone – connectivity, communications, and context, for instance – are so far

Figure 7-6. The classic mobile game, Snake (reproduced by permission of © Nokia).

rare. An example is *Before Crisis: Final Fantasy VII*, a network-based action RPG, developed and published by Square Enix, Co., Ltd. for the NTT DoCoMo FOMA 900 series. This is the first game of its kind, developed exclusively for mobile phones. Among other features, the game uses the built-in camera, and allows networked cooperative play.

One research project particularly targeting mobile multiplayer games is the Multi-User Publishing Environment (MUPE)[27] (Suomela et al. 2004), developed in the Nokia Research Centre and provided for developers under Nokia Open Source License (NOKOS 1.0a).

MUPE is an application platform for mobile multi-user context-aware services, allowing anyone to create their own mobile services, games, and applications. To achieve these goals, MUPE uses a plethora of technologies to reduce the complexity of developing mobile services in client and server sides. From an end-user perspective, MUPE is a collection of services, and a service browser. Users can browse with MUPE client for new services, connect to them, and create persistent user accounts. After this, they can use many online services, such as games, messengers, and so forth. A single MUPE client installation is enough for all functionality, since once installed, MUPE updates itself as necessary. MUPE is used in two ways. The services are used by installing the MUPE client, and the services are hosted, and developed, by installing the back-end packages.

---

[27] http://www.mupe.net/

There are some issues related to mobile games. Since the games developed for desktop computers and settop-box game consoles are played next to the device, it is assumed that every player has a device available, and they have purchased a licence to play the game on that device. On the contrary, with mobile games, the *ad hoc* nature of game play implies that potentially any person may participate on a multi-player game *in situ*. Now the question is: do all the players need to own the game, or a licence to play?

The question of licence ownership is sometimes related to, surprisingly, personal content created while played. In case some game-related content (metagame elements) is retained after the game is over, the player needs to own a licence, whereas in other cases only one player may own the game; they then host the game and share the clients that are needed (for instance, Game Boy Advance). These approaches can also be combined: in the beginning, one can play without a licence, but purchasing the licence enables creating and retaining the metagame elements.

7.4.2 Personal Content Types Related to Games

From the personal content point of view, user-created content related to games can be divided into two. Some content is created *implicitly*, while playing, such as high scores, whereas some content is created *explicitly*, such as game characters. We will discuss both of these types below.[28]

The simplest form of personal content related to games is *score*. Traditionally, scores have been stored locally as a top-ten high-score table, accessible only through the game software. However, these days many games offer the players a way of posting their scores on the web site, thus producing a potentially global high score table. On the gaming scene, game scores are the equivalent of stock valuations – important quantities that reflect a gamer's achievement, coolness, and prestige.

In addition to achieved scores, the current situation in a game can also be stored. This is referred to as a *saved game*. In the most simple form, the game is saved locally on a mass storage device, where it can be later retrieved, allowing the player to continue gameplay from the last saved position. There are two basic approaches to saving games: either the player is free to save the game at any time, or alternatively,

---

[28] Please note that in most domains the same division between explicit and implicit content creation applies. However, with games it is perhaps more clear than in others (Chapter 3).

saving is only allowed at certain points as the game progresses. Some games allow sharing saved games. Nevertheless, with most of today's games often being extremely complex and time-consuming, saving the games is a necessity. In this respect, saved games are a valuable form of personal content – many hours of playing may be lost in cases where a saved game is corrupted or lost.

Game movies are recorded live gaming sessions. There are two options for creating game movies: either an external screen capture application is used, or the game has built-in functionality to record events, effectively providing an accurate reproduction of the played game. Furthermore, some games, such as *Sims*, provide only a still image capturing capability for recording.

There are many uses for game movies. Obviously, they act as proof of achievement. Another application for game movies is *ghosting*, often exploited in car racing games. In ghosting, a previous game session of a player is displayed on the screen as a ghost car, allowing guiding or competing against oneself.

Closely related to saved games and game movies are *walkthroughs*. They are accurate textual documents, often accompanied with screenshots, explaining the actions and tricks needed to complete a section of a game, or the whole game. Game movies are one form of providing these guides.

An interesting phenomenon related to game movies is *machinima*, which refers to using the game engine for making movies.[29] In other words, it is about shooting films in virtual reality. Machinima has become immensely popular during the past few years. The movies vary from short clips to nearly full-length movies, from science fiction and horror titles to funny stories about the games the film was developed with.

Somewhere in between implicit and explicit content, lie *in-game messages*, messages sent to other players from within the game, while playing the game. In-game messages are obviously explicitly created, yet their relation to the game varies. Nevertheless, messages add to the social aspect of multi-player games.

Many games allow the user to define their play *character*, that is, its personality, strengths, and weaknesses. A game character is a step towards explicitly created game content. Similarly to characters, other game *items* can be created. A recent example of crafting game characters and game items can be found in *World of Warcraft*. However, user-created content has played an important role in much earlier

---

[29] http://www.machinima.com

games. For instance, early MUDs[30] allowed the players to craft their own in-game elements, such as areas, monsters, and quests.

For some games, such as *Second Life*,[31] the whole business model is built around user-created game content. As stated in the game home page:

> You'll also be surrounded by the Creations of your fellow residents. Because residents retain the rights to their digital creations, they can buy, sell, and trade with other residents. The Marketplace currently supports millions of US dollars in monthly transactions. This commerce is handled with the in-world currency, the Linden dollar, which can be converted to US dollars at several thriving online currency exchanges.

Essentially, in *Second Life* you are able to make real money in a virtual world by selling virtual assets, that is, user-created digital personal content. Virtual asset trade is indeed a trend worth keeping track of.

7.4.3 Modding

By far the most common activity related to explicitly-created personal content is a phenomenon known as *modding*. Modding is a participatory practice where "game fans modify and extend officially released game titles with their own creations" (Laukkanen 2005). The modifications (mods) created by the players range from skins (graphics that alter the appearance of some game elements) to total conversions (new games that are built on existing game engines, including custom graphics, custom audio, and custom executable code).

The enormous success and maturity of modding has not been overlooked by commercial game developers, who intentionally support modding by creating mod-friendly titles. Furthermore, many developers currently consider user-created content an important part of their marketing strategy. This results in the inclusion of modding tools in the game sales packages, developer-hosted support forums, and modding contents organized by the developers.

Skinning, referring to altering the visual appearance of a game, is the simplest form of modding. There are two approaches for skinning: either all graphics are created from scratch (including textures), or the existing graphical elements are configured. Skinning from scratch requires a considerable amount of work.[32]

---

[30] Multi-User Dungeon; usually a text-driven online computer game that includes role playing, fighting with enemies, and social elements.
[31] http://secondlife.com/
[32] Besides games, many other kinds of applications, such as media players, also allow skinning.

In between skins and total conversions, there are numerous forms of user-created modded content available. Typical examples for single-player games are levels, the definition of tasks that the player needs to complete before advancing to the next level, complete with environmental features. Likewise, in multi-player strategy games, user-created maps are often used.

The computer game that arguably transferred the modding scene from a niche activity to mainstream was *Doom* by id Software in 1993. Additional tools allowed the users to produce their own levels, skins, and total conversions. *Doom*'s success can be considered to be greatly boosted by the rapid distribution of these mods through the Internet.

Fan fiction as such is not a new phenomenon. For ages, media fans have produced their own creations based on their favourite books, movies, or TV shows. What is new in modding is the digitalization of both the content and the distribution channel, making it a global practice that is evolving rapidly.

In a larger context, modding is an example of a participatory culture (Jenkins 1992), where the audience takes an active role in the circulation of media texts, blurring the boundaries between the audience and the content producers. This is closely related to citizen journalism (section 3.4).

As far as the actions performed on modded content are considered, most GEMS phases can be identified. Mods are created or downloaded, and then enjoyed, either by playing the game, or by using them as a basis for further modifications. Finally, sharing mods is a common practice – in fact, sharing is the major motivation for creating them in the first place.

7.4.4 Discussion

What is worth noticing is the fact that unlike some other user-created content, there is often real money involved in game-related content creation. The fact that the user *pays* for the opportunity to create content is an interesting notion. Compare this to paying for each photo you take with your camphone!

CRAP DE LA CRAP

Eddie's an on-off gamer. He likes to try out new game warez, but he's not deep into it like some of the boyz in the 'hood. However, once he decides to get busy with a new game, he will really dig deep into it and only come out when he's exhausted every possible twist that game has in store for him.

It's then ever more frustrating when the game fails on him. If he decides to get friendly with a piece of code, it better be trustworthy. Not like Surreal Shards, that *crap de la crap*. He had invested two weeks in decoding the access to the time portal, spent countless hours talking to other people in that MMORPG, finding his way through the needlessly complicated social order of the game, and even invested some real money (borrowed from his sister) to rent the EPS powers. All of this, in vain.

Last night, when he logged onto Shard's site again, he was greeted with an error message boasting "You don't exist. Go away." After some trying and cursing and more trying, he then gathered from the site's chat forum that this did indeed happen every now and then, and the only way to get over it was to start from scratch. Eddie could not believe it. How is it possible that an online game could lose a gamer's profile, so that all that work invested in the game would be lost? How on Virtual Earth could anyone in their sane minds allow this? Who could he sue?

After a few days of useless complaints on the server's customer service, Eddie calls it quits. They don't want his business, he's out. No more Shards for him. Another case of Never Trust a Piece of Code. Especially, emm, this pirated game that he copied from a friend.

In addition to the content types discussed above, mobile games bring some new elements to the game. As discussed in section 7.2., location is an inherent feature of mobility. Not surprisingly then, mobile games location is, or will be, a key element both in game design and in game-related content. Customizations based on location, built on top of the generic game data, will play an important role. Games that systematically blur and break the boundaries of the game in spatial, temporal, and social dimensions, are referred to as *pervasive games* (Montola 2005).

For location-based customization, there are two approaches. Either the players make the customization – and the required tools – themselves, or there will be service providers that deliver either the customization or the tool, or both. The role of the tools should not be underestimated, as is the case in mods today, where the game developers provide modding tools themselves and support modding explicitly, the customization tools needed for location-aware games are becoming increasingly important.

What is interesting in the relationship between games and personal content is that in many cases the content is not an end in itself, but rather an indirect vehicle for experiencing the actual content – which is playing and enjoying the game, and participating in the community. In other words, even though game-related content plays an important role, it still has a secondary role compared to many other application areas that explicitly deal with content.

There are risks involved in games, as with any type of software. Especially, virus attacks via games are a potential threat – many games allow versatile connections between players, thus providing paths for viruses to sneak in. Also mods may contain malware such as viruses, worms, and Trojan horses. Not to mention illegal sharing of copyrighted games – most file-sharing networks contain loads of cracked games that, when downloaded, do not arrive alone.

Another less explicit, yet equally potential, game-induced risk is related to in-game messages. The communication channels discussed above often provide for the connected gamers to communicate to each other. As with any communication method, in-game messaging is subject to abuse and real-life risks such as bullying and predators.

An interesting aspect related to mods is censorship and, often closely related, copyright. In some countries, the law may require that the content is checked before it is released to the public. Likewise, online games include screening for in-game messaging, character names and appearance, and so forth. The moderator may replace words they find unacceptable with graphical symbols, such as # and %. Mods are also screened before they are released. Obviously, it is impossible to extend these requirements to mods built for stand-alone single-player games, or to those that are distributed unofficially. A notorious case is a mod called *Hot Coffee* created for *Grand Theft Auto: San Andreas*.[33] The mod enabled a hidden "Adults Only" mini-game within the game. As a result, several stores pulled the game off the shelves, and in some countries the game rating was changed. A lawsuit was also filed.[34]

Speaking of unofficial mods and DRM issues, the relationship between creating mods and distributing them is unclear. In cases where the game developer has not released any tools or documentation for creating mods, reverse engineering is needed to be able to create modifications; this can be considered cracking the game. Even though, such activity can be considered dubious from the game developer's point of view, as what is clearly illegal is the distribution of such mods.

Another example related to copyright deals with character appearance. In an MMORPG title *City of Heroes*, where the users are able to configure the character appearance, the developers made some characters appear like cartoon characters from Marvel comic books, such as *The Hulk* and *X-Men*. Marvel Enterprises consequently sued the

---

[33] http://www.gamespot.com/news/6129500.html
[34] http://www.gamespot.com/pc/action/gta4/news.html?sid=6143276

game developer for breaking copyright law.[35] Furthermore, in the *World of Warcraft*, the characters are not allowed to have names similar to those appearing in the game.

With personal content, privacy may always be compromised, and content relating to games is no exception. The game data – game histories, saved games, game characters – may be sensitive. Their monetary value may be considerable, as in the case of the online game *Second Life*, discussed earlier. In eBay,[36] a popular Internet auction site, there are game characters and valuable items offered for bidding. Such virtual economy has rapidly developed around online games, and a considerable amount of money is involved.

Habbo Hotel[37] is a similar web service, targeted at youngsters and offering chat, games, and other entertainment using a hotel-like metaphor. *Habbo Hotel* can be considered a kind of virtual doll's house: the idea is to develop your own virtual characters, and to decorate your own room. *Habbo Hotel* uses *habbo coins*, virtual money that can be purchased with real money. Habbo coins can then be used to enhance the experience, for instance decorating the room with virtual furniture, or buying a virtual pet animal.

To summarize, games are an important domain, both from the business perspective and from a personal content point of view. The development in the gaming domain is particularly fast, and one may well expect to see some novel forms of GEMSing personal content appearing from within the playing community.

7.5   Other Domains

So far we have discussed obvious domains of personal content; these are domains familiar to many researchers and practitioners in academia and industry. Next, we introduce a few less featured domains where personal content has a role to play. In domains such as these the importance of personal content may be boosted rapidly and unexpectedly in the near future.

7.5.1 Personal Training

To keep in mind that personal content truly lives not only on traditional computer-based domains, but also on consumer devices that are not

---

[35] http://www.usatoday.com/tech/news/2004-11-11-marvel-sues-over-avatars_x.htm
[36] http://www.ebay.com/
[37] http://www.habbo.com/

Figure 7-7. Suunto t6 wristop computer and heart rate measuring belt (Photo: courtesy of Tero Hakala).

usually considered computers, we have chosen the first case. We still wish to stress the fact that even though these devices do contain embedded computing hardware, they are not considered to be computers by general public.

Personal training is a representative application area that illustrates our point. Personal fitness data (or content) is created with various fitness devices, such as the Suunto t6 wristop computer (Figure 7-7). The device consists of two pieces, a wristwatch main unit, and a heart rate measuring belt that is worn around the chest.

In addition to standard time and date functions and heart rate measurement, the device is also capable of measuring temperature and air pressure. Pressure can be used to estimate the altitude. With some extra accessories, such as a foot pad (an acceleration sensor that is attached to a shoelace and can measure the speed by analyzing the foot movement patterns), distance and speed can be calculated. The same can be achieved with a GPS receiver, such as in Suunto X9i.

From a personal content point of view, we are especially interested in the heart rate measurement and travelled route with GPS-enabled devices (section 7.2).

The main functionality in any device targeted at personal training is simple enough. The device connects wirelessly to the heart rate belt worn around the chest, and once the "Start timer" button is pushed, it starts recording the heart rate at defined intervals, effectively creating a log, until the "Stop timer" is pressed. Then, the device can typically show the average heart rate, the estimated calorie consumption, and minimum and maximum heart rates. Often, this information is also available in real-time while exercising. These are typical dedicated devices that focus on "G" actions.

However, many personal training devices are capable of doing considerably more than this. Suunto t6, for instance, can connect to a PC via USB. The logs can then be transferred to the PC for further analysis. From the information recorded by the wristop computer, the PC software can analyse excess post-exercise oxygen consumption (EPOC), heart rate, respiration rate, ventilation volume, oxygen uptake, energy consumption, and training effect. This information can be further used to make training plans and programs for optimal results related to desired training goals, such as losing weight or training for a forthcoming marathon event.

Other features include presenting the training information in graphical form, and backing up the content. An additional dimension for Suunto is its online community support forum,[38] which allows people to discuss, compare, and share content.

Besides general exercising, there are also dedicated devices targeted at certain types of sports, such as diving and golf. These devices are capable of recording and analyzing information specific to the sport in question.

So far, personal fitness data has not been regarded as personal content. Many devices have been unable to export the data, people have not bothered to store the data, and there have not been easy analysis tools available. All this is changing and we believe that, in the next few years, GEMSing personal fitness data will become an important application area for mobile devices in general.

7.5.2 Movie Subtitles

Another example of personal content, which is not usually considered in the domain of personal content management, comes from the movie world: subtitles for (DVD) movies. Typically, movies are delivered on tangible DVD disks, complete with menus, soundtracks (perhaps several version for dubbed languages), extras, and subtitles for regions that prefer some visual disturbance on the screen to losing the original voices of the actors.

However, today movies are also delivered via other means, such as buying on a netstore, or as illegal copies from file sharing networks. These movies are in many cases delivered without any extras, so as to keep the bandwidth requirements to a minimum. As a result, subtitles may be missing.

For many, missing subtitles may not be a problem, but if the viewer does not know the original language, they are in trouble. Not surpris-

---

[38] http://www.suuntosports.com/

ingly, users have started to deliver the subtitles themselves. Many current video players are capable of reading the specific subtitle files and overlaying the subtitles over the video content while the video is being watched. In essence, subtitle information consists of text strings (the actual translations), and the frames between which the associated text should remain visible while watching the movie.

Subtitles are an excellent example of personal content that has been derived from existing content and transcoded to another content format. In addition, a lot of creativity is involved in the translation phase. Subtitling is also a social process, as often the subtitling of a movie has been split for several persons, such as to ease the workload and, in many cases, proofreading is involved. Not surprisingly, vivid communities have evolved around subtitling movies, such as DivX Finland,[39] founded in 2003, and with more than 9 000 members (as of November, 2006). The community has its own archives and policies regarding the processes of translation and proofreading.

The GEMS actions for subtitles are much the same as with any other digital content. People search, download, edit, manage, and share subtitle files as they would do with any text information. The difference is that the subtitles are usually not useful without the movies they refer to. In a way, subtitles are then metacontent of movies.

7.5.3 Flash, Comics, Animations

After the birth of the Web, bedroom artists have had dramatically better chances of publishing their work. Poets, writers, and musicians can now easily upload the fruits of their labour to numerous web sites. One area that has gained public attention is with comics – *webcomics*, to be more accurate. There are thousands of online comics being published,[40] most of which would have had little chance of even a minor circulation in the paper-based era. This is a kind of content that wavers on the border of public and personal: on one hand, it is targeted for exposure; on the other, it is usually not meant for profit but as a vehicle of self-expression.

Animations of various kinds, using a multitude of technologies, are also popular. Most of the animated material on the Web uses Flash[41] as the presentation tool. It allows rich multimedia presentations that can also respond to user actions. It is the key format for web advertise-

---

[39] http://www.divxfinland.org/ (In Finnish)
[40] http://www.webcomics.com
[41] Adobe Flash Player, formerly made by Macromedia: http://www.adobe.com/products/flashplayer/

ments, but also attracts non-commercial multimedia developers. As a result, large amounts of Flash content are available on the Web,[42] ranging from slide shows to click-through novels, and from music video mashups to on-line arcade games. Again, the content is mostly contributed from personal points of view but targeted at web viewers.

7.5.4 Discussion

The above discussion illustrates the countless kinds of personal content out there. Moreover, as the Web keeps mutating and the Long Tail phenomenon strengthens, several new innovative content types will have been born between the time we wrote this book and you get to read it. This is all the better – it is amazing to see what people can come up with once the means are available. There is no limit to creativity when it comes to personal content, as long as the technology does not get in the way. We hope this book helps to avoid some of the technology limitations, or at least to be aware of them.

7.6 References

Ashbrook D. and Starner T. (2002) Learning significant locations and predicting user movement with GPS. *Proceedings of the Sixth International Symposium on Wearable Computers* (ISWC 2002), pp. 101–108.

Blades M. and Spencer C. (1987) How do people use maps to navigate through the world? *Cartographica*, Volume 24, Number 3, pp. 64–75.

Boyd D. (2006) Identity Production in a Networked Culture: Why Youth Heart MySpace. *American Association for the Advancement of Science*, St. Louis, MO. Feb. 19.

Davis M., King S., Good N., and Sarvas R. (2004) From context to content: leveraging context to infer media metadata. *Proceedings of the 12th annual ACM international conference on Multimedia*, pp. 188–195, ACM Press, New York, USA.

Gartner (2006) *Forecast: GPS in Mobile Devices, Worldwide, 2004–2010.* Abstract available at http://www.gartner.com/DisplayDocument?doc_cd=144746

Getting I.A. (1993) The global positioning system. *IEEE Spectrum*, Volume 30, Number 12, Dec., pp. 36–38 and 43–47.

HGVNU (2006) Harrison Group / VNU Teen Trend Report, 2006. Summary available in http://www.harrisongroupinc.com/index.php?q=node/57 (Last checked Dec. 8, 2006).

---

[42] Some examples here: http://www.flashgallery.co.uk/

Hyman P. (2005) State of the industry: Mobile Games. *Game Developer*, Sept., pp. 11–16.

Jenkins, H. (1992). Textual Poachers. *Television Fans and Participatory Culture*. Routledge, New York and London.

Laukkanen T. (2005) *Modding scenes: Introduction to user-created content in computer games*. Online document, available at http://tampub.uta.fi/tup/951-44-6448-6.pdf

Lehikoinen J.T. and Kaikkonen A. (2006) PePe field study: constructing meanings for locations in the context of mobile presence. *Proceedings of the 8th conference on Human-computer interaction with mobile devices and services*, pp. 53–60, ACM Press, New York, USA.

MacEachren A.M. (1995) *How Maps Work*. The Guilford Press, New York, USA.

Montola M. (2005) Exploring the Edge of the Magic Circle. Defining Pervasive Games. *DAC 2005 conference*, Dec. 1–3. IT University of Copenhagen. Available online at http://www.iki.fi/montola/exploringtheedge.pdf

Naaman M., Harada S., Wang Q.Y., Garcia-Molina H., and Paepcke A. (2004) Context data in geo-referenced digital photo collections. *Proceedings of the 12th annual ACM international conference on Multimedia*, pp. 196–203, ACM Press, New York, USA.

Sarvas R., Herrarte E., Wilhelm A., and Davis M. (2004) Metadata creation system for mobile images. *Proceedings of the 2nd international conference on Mobile systems, applications, and services*, pp. 36–48, ACM Press, New York, USA.

Schwartz E. (2006) The Net Finds its level. *InfoWorld*, Oct. 23, Issue 43, 10 p.

Suomela R., Räsänen E., Koivisto A., and Mattila J. (2004) Open-Source Game Development with the Multi-User Publishing Environment (MUPE) Application Platform. *Proceedings of the Third International Conference on Entertainment Computing*, pp. 308–320. Springer.

Toyama K., Logan R., and Roseway A. (2003) Geographic location tags on digital images. *Proceedings of the eleventh ACM international conference on Multimedia*, pp. 156–166, ACM Press, New York, USA.

Chapter 8: Timeshifting Life

application context data information

life **metadata** object

ontology people sharing user vision

"The future is here. It's just not widely distributed yet."

William Gibson

The above quote contains inherent truths about any research and development related to new technologies. Based on the current state of technology, we make assumptions (visions, if you prefer) on how things will be, and then seek means of reaching that state of affairs. It is a matter of prior research, societal developments, business decisions, and other more or less controlled factors that finally determine who gets the vision right. In most cases, the big picture will only reveal itself after some time has passed, and in hindsight it is easy to see what really happened. We will see that many of us were right – and wrong.

The combination of these fragments of successful visions can be called the future of the domain in question. And many parts of this future exist today, scattered across laboratories and start-up companies around the globe.

Personal Content Experience: Managing Digital Life in the Mobile Age   J. Lehikoinen, A. Aaltonen, P. Huuskonen and I. Salminen © 2007 John Wiley & Sons, Ltd

This book has largely concentrated on the *status quo* of managing and experiencing personal content. This chapter extrapolates some of the issues previously discussed, to paint a vision of the future of personal content. We are not expecting to get it exactly right, as our crystal ball warns us that there are many changes ahead. Still, most of these issues will be crucially important to the area of personal content. Here is a summary of what we foresee:

- Personal content will become "smart", that is, aware of its semantics, and get connected to other pieces of content; the origins of a content object will be traceable.

- Personal content applications will become interoperable through metadata.

- Successful metadata machinery will remain invisible to users in most situations; smart UIs will be metadata-aware.

- On-line societies will be crucial for creating, managing, and experiencing content – sharing to the extreme, peer-support (recommendations, answers), metadata created by social interaction.

- Automated functions will be powerful in limited domains.

- Metadata magic will rise from machine cognition, data mining, context awareness, ontologies, folksonomies, object decomposition, metadata extraction.

- Fake metadata and privacy will be key concerns.

- P2P networks will graduate from file sharing to metadata sharing.

We will now look at these issues in more detail.

8.1   Metadata in the Years to Come

Throughout this book, we have stressed the importance of metadata as a key technology that is needed for preventing the personal content boom to explode in our faces. There are some potential directions that metadata-related technologies will evolve. The following factors describe those directions:

Drivers: the explosive growth of digital content will drive the need for metadata. Mere file attributes, such as name and date, do not sufficiently describe its contents and cannot reveal the fine-grained differences of the content objects that are (partially) created and edited by other persons. Moreover, automated handling of content will require machine-readable descriptions.

- Machine-readable metadata is needed for automation

Metadata creation: the trend is towards automatic metadata creation – especially when manual annotation does not pay. Online communities will (semi-automatically) annotate content through tagging and ontologies, in cases of content that is meaningful to and desired by a large number of users. Interactions between people in general will turn out to be important sources of metadata. One major metadata creator will be the mass-market industry, spurning out huge amounts of metadata via product information, advertisements, services, and so forth.

- Communities will be a key source of metadata

Metadata usage: basic metadata will be commonplace, being adapted to people's everyday lives. Raw metadata will be mostly invisible to the users, yet it will be everywhere, deeply embedded in applications, operating systems, data, and physical objects. Occasionally people will see glimpses of metadata as the tip of the iceberg, when automatic rating or summarization factors are presented to the user, for example. Metadata will be multi-form and used in unexpected novel ways. Metadata support becomes a competitive "must have" for applications. Application interworking will be the norm, not the exception.

- Good metadata is invisible!

Technology: The Semantic Web will make breakthroughs within limited domains. Generic ontologies will still be largely just dreams. Tags and folksonomies will run into complexity and scale-up problems. Decomposition of media items will become commonplace: each individual media fragment can be uniquely addressed, re-used, and referenced. Automatic metadata extraction, with various techniques over multiple content types, becomes the norm.

- Upcoming tech: domain-specific ontologies, folksonomies, decomposable content, metadata extraction

Risks: compromised privacy and bogus metadata will turn out to be the largest initial concerns. Automated metadata extraction from content will improve, but will still guess wrong in cases of more difficult content. Non-compatible metadata formats will create a need for translators, but even that does not guarantee interoperability. Deliberately faked metadata or forged ontologies will destroy many schemes based on open contributions. Moreover, knowledge acquisition will prove too hard for most real-world problems. If the metadata cannot be captured as a by-product of existing applications, it will remain uncovered.

- Fake metadata and privacy are key concerns

It is of crucial importance to start attaching metadata to content now. The more we have standardized annotations and associations available, the less likely we will have unorganized personal content floating around in the future – that is, content beyond most GEMS actions. The question is: how will this take place?

8.1.1 Metadata Enablers

We have stressed the importance of automatically created metadata. Several related existing technologies that are currently under heavy research can be identified in such fields as image analysis, language understanding, sensor processing, context awareness, pattern recognition, and data mining in general.

- Metadata magic comes from machine cognition, data mining, and context awareness

We could consider a brute force approach, where metadata is abstracted from raw data, once again regardless of content type. A typical example might be scanning through all content, at the binary level, and identifying ASCII or Unicode strings that make up human-readable words. These words are potential candidates for describing the file in some aspect. However, there will be plenty of noise in the results, so consistency and integrity checking will be needed. Besides, much of the interesting content today comes in non-textual forms.

Metadata should be streamlined not only across applications, but also across content types. Metadata-based similarity recognition across different media formats could lead us to automatic cross-media categorization. Once metadata is created for a certain content object, other related pieces should benefit from the same metadata, regardless of the content type.

- Metadata will propagate between related content items and applications

Various kinds of personalization and adaptation are needed when using metadata from multiple sources. For instance, we should have a method for "normalizing" the values that others provide to my own personal scale: your one star song rating is actually my 1.8 stars.

- Metadata semantics vary between people

Content-based retrieval (CBR) is a promising technique for searching content (section 5.2). In CBR, the content itself is analyzed in order to identify objects or other features. For instance, the visual characteristics of a photo can be analyzed and checked for people, objects, and weather. The identified information is then stored as metadata, in textual or other form, to allow rapid machine operations. As usual, any successful content recognition will not function universally, but will be domain- or context-specific.

- Content-based retrieval is powerful in limited domains

The roadmap towards more advanced content-based retrieval techniques is incremental. That is, some unsophisticated analyses can be performed first, say, "this photo was shot at night," and once the methods become more advanced, the created metadata can be augmented with additional information. Context-awareness research will bring in several enablers.

Online communities are crucial for metadata creation. Therefore, technologies that support easy creation, verification, and sharing of not only content, but also metadata, are of specific importance. As an example, we envision a "Whole Earth" distributed catalogue, where a P2P network recognizes and annotates content. Furthermore, intelligent metadata sharing could be enabled by global author identification and personal global metadata namespaces (RDF). It is also likely that we will have good and solid techniques for making the globally shared metadata anonymous, or verinymous, depending on the situation.

- P2P indexes world's content

However, communities may produce more than just mere metadata. Communities may be harnessed to provide entire ontologies on specific domains, especially once the payoff is high enough. Even today, user-contributed ontologies provide semantics for restricted domains (for instance, the dmoz open directory project (ODP)[1] had

---

[1] http://www.dmoz.org/

over 4 million sites and over 590 000 categories as of August 2006). The same significant growth has taken place in other user-contributed projects, such as freedb,[2] and may become an important factor in defining universal ontologies.

- Open ontologies needed for scale-up

What is essential in this respect is the interoperability between different ontologies. This is a huge challenge, and it is likely that the ontologies will remain incompatible for several years to come, primarily due to their anarchistic nature of evolution and the inherent difficulty of knowledge representation.

SHOOT THE BASS PLAYER

Eddie took Angie to see The Kinkees' gig in The Hard Rat Hall. She's not really into their retro-punk music but, having listened through Eddie's playlists dozens of times, she knows most of the songs. The gig is great, the band seem to be really wound up, the audience is packed and jumping up and down. As usual everybody is filming and bootlegging the band, as they explicitly invite contributions to enhance their OuterSpace site.

The bassist is quite cool, Angie thinks. He's not your dad's bass player who would stand quietly in the corner. No, this guy is up there, flying those silly rock jumps in sync with the lead guitarist. Kind of cute, too. Who is that guy? Why does he look so familiar?

Angie works her way to the side of the stage. The speakers are pumping air so that her blonde hair is flown in her face. Loud, but fun. She gets just next to the bassist, he sees her waving and poses for her in that rockstar fashion while she snaps an image. Wow, I'm a real fan now, Angie muses.

Angie uploads the image to RokkenRoll, the prominent rock music trivia archive. Angie sees that the site features live footage from tonight's show. Looks like a number of people there were uploading too. Angie queries the service for matching images of the bass player. It turns out there are hundreds of matches – this guy must be famous – so she limits the search by the context information of the event that other people have already fed in. That's better, about 15 matches.

One of the images has already been tagged with the bassist's id: Roy Kee. Weird name. The id is a live link, which Angie saves for later perusal. She could also chat with other audience members and locally share footage of the gig, but not now, thanks. Now it's time to find her way back to Eddie, who's somewhere there in the middle of that cheering and waving crowd. She jumps in to dance. This music actually rocks after all . . .

---

[2] http://www.freedb.org/

Loads of bulk metadata is today being created automatically, as RSS feeds accompanying new web content (such as photoblogs). We expect this trend to expand and continue, and propose that RSS integration to metadata creation schemes could easily pay off in the future.

- RSS is a major metadata source

An interesting notion related to the smartness of metadata is the awareness of its origins. In other words, we could develop schemes that allow tracing a content object to its roots in the case of, for instance, multiple generations of copies. A potential yet largely unstudied enabler is the usage of *genetic coding* for producing globally unique identifiers (GUID) for content objects. Such an identifier would allow us to check the ancestor object of any content object, which again allows us to check its ancestor, and so forth. Undoubtedly, there are several unsolved questions and potential threats related to such traceability.

- Tracing objects back to their "roots"

One more example of enablers is smarter user interface components that have built-in support for using metadata. For instance, list and grid components could provide sorting, filtering, and grouping properties, where the user could change the organization criteria quickly. This would facilitate quick ways to organize content and generate different kinds of views to the content mass (see the discussion on Matrix Browser in Chapter 6). Another example is a zoomable user interface (ZUI), where the widest zoom level uses automatic grouping, sorting, and summarization, and as the users zooms closer, new and relevant information related to the objects is revealed (Figure 8-1). Furthermore, metadata enables dynamic browsing of content based on associations and relations between the content objects, instead of file-folder hierarchies.

- Smarter UIs have metadata functionality built in

8.2   Metadata Creation: Top-Down or Bottom-Up?

On a higher level, two approaches to smart metadata can be identified. Either the *top-down* approach dominates, where metadata is included

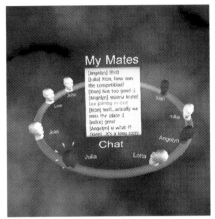

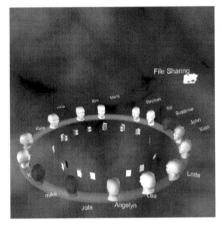

Figure 8-1. A zoomable UI concept focusing on people and content.

in well-defined formats and is controlled by global standards and metadata creation tools, that can make consistency checks and rely on strict universal ontologies; or the *bottom-up* approach becomes dominant, where metadata is created by individuals, in non-standard formats, no ontologies are in use, and the error rate is high.

In Chapter 4 we claimed that as far as consumers are concerned, manual metadata equals to non-existing metadata in many cases. Therefore, the metadata creation is deeply connected to the level of its automation. It is safe to expect that automatic metadata entry will increase rapidly – provided that the whole personal content ecosystem develops to a mature enough state – and will form the base of all existing metadata. This trend is clear as of writing. The inclusion of automatic metadata beyond mere date and time is increasing as we speak, and is especially visible in online music stores, such as Apple iTunes.[3] The information about artists and songs is automatically included in all purchased music.

The options given above – top-down or bottom-up – will probably co-exist for some time, so that the dominance makes a slight transition from bottom-up to top-down. The reasons are simple, as the metadata is increasingly created automatically today, without any universal *de facto* standard, there will be loads of incompatible, media type-, application-, and manufacturer-specific metadata formats around for some time. Also, if people create metadata manually, the semantics of an attribute and its value may differ between people.

Once new applications and devices supporting more generic standards of metadata management become commonplace (devices that are capable of converting proprietary metadata formats to the then-available *de facto* standard), the dominance is shifted towards the top-down approach. Pre-existing metadata is converted and checked for consistency once it is encountered, which slowly converts old metadata to new standards.

As discussed in Chapter 5, extensibility of the *de facto* metadata standard allows any third-party content provider to implement their own filters and converters for any file type, make their contribution to the global ontology, and thus making the rare content type future-proof. Nevertheless, the two approaches to metadata creation will continue to co-exist now and in the foreseeable future.

One should not underestimate the influence of communities related to personal content and metadata creation. For instance, in Chapter 7 we presented YouTube[4] and similar social services and applications that allow creating and sharing content and metadata. We again emphasise that virtual communities related to personal content are a strong candidate for driving metadata creation and subsequent metadata sharing.

8.3   Show Me the Money

Provided that the importance of metadata is as predicted above, it seems evident that metadata can enhance existing applications and services in a number of ways. This better, stronger, faster functionality translates to a better user experience, which means happier customers, which means more money.

---

[3] http://www.apple.com/itunes/store/
[4] http://www.youtube.com/

Some potential new business domains can be envisioned. For instance, we will see an increasing need for services that provide automatic metadata tagging by, for instance, detecting objects in an image, or identifying songs based on a short sample. Another potential metadata-related business opportunity is related to cleaning services that aid in compressing or removing old versions, obsolete, or false content objects and their relations and metadata.

There will be people who sell their own context data, connected with and related to metadata that is relevant to those contexts. Getting contextual data from people in real-life situations allows content aggregators to find out what content is relevant to those situations. On the other hand, content collected by crowds may be indexed on the basis of the context obtained from these "context jockeys". The context then ends up in the metadata of the content items.

SHE'S A ZERO

Eddie's so disappointed with Angie. All of a sudden, his shiny smiley dream babe walks in and, without warning, matter-of-factly informs him that she's got herself a jockey account.

At first, he can't believe his ears. A jockey cellphone deal? Selling all her data to the devil!

"You what?" Eddie started.

"Heard me, hon. With Zero. Free voice, text, video, IM, sharing, TV, whatever."

Angie gets a blank stare from Eddie. After all his preaching and patient explaining, his girl still does not get it. There's no such thing as free access – it comes with a price.

"Look, we've been through this before. It's free because they get all your usage data! They'll know everything about you!"

"So what?"

"But you're being abused! Exploited! You're selling your –"

"Aw c'mon. To hell with that. Everybody's selling their context. It's like, you know, the 21st century! Get real."

What really makes Eddie mad is that she's right. There are loads of people on the street who couldn't care less about their privacy, and opt instead to openly offer their precious context profiles, location data, music use, purchase details, anything. All that in return for some fast bit pipes and a few dismal services. They don't care. That's so wrong!

"Angie darling, it's not just you, but you're stocked with stuff about me, get it? Anything you do with me will contain bits and pieces of my data too."

"So?"

"I don't want that! You're not only exposing your sorry ass, but mine too!"

Angie forces that wry smile that she knows he hates, raises her eyebrows mockingly. "Aha, hacker boy, you're scared. 'fraid your downloads will end up in the hands of' FAM."

That was shorthand for UNFAM, the United Networks Fair Media Agency.

"Bloody well right I am. You know we're loaded with illegal content. You are, too, by the way."

"But everybody's doing that!"

"You stupid cow! Everybody's doing that but not shouting about it to the police!"

"Yes they are!"

"No, they're not! Nobody with half a brain!"

"They are too! All the girls in the 'hood are getting into Zero. They don't care. Data wants to be free, you know."

Eddie can't believe this. His girl, after all these years, is turning into a context jockey. A Zero. A zombie. Braindead. After all their precious moments that they have kept to each other. He can't have this.

"Look, Angie, it's taken me years to collect all that content."

"Yeah. But you, it cost you nothing. Not a nickel."

"That's irrelevant. I spent a lifetime hunting for it. And then you go and expose all of our stuff. Ours, baby. Yours and mine."

Angie hesitates. The lava lamp on Eddie's table pops a blob. Then she offers, "Maybe I don't need that stuff anymore."

Ok, this is it. Eddie has been afraid of this for some time. They're falling apart. Been distant for weeks, now getting even further away.

What is probably more important than direct revenues from metadata is that access to metadata-related server infrastructure (such as ontologies and content catalogues) may be necessary for total product offerings, not only as a service business as such. Presumably support for metadata and standardized GEMS actions in content-related applications becomes first a key differentiating factor, and application providers will fast adapt once the general public turns towards products they trust and know are able to handle their memories and digital life for decades to come. However, in the long run, such support cannot alone positively differentiate applications. We assume that metadata-enabled application interworking will be a key competitive differentiator as far as content-related applications are concerned.

Recent buy-outs of social and content-related web sites, such as MySpace and YouTube, are an indicator of the increasing importance of peer support and communities within personal content experience. The revenue models may still be unclear, but the potential consumer base for targeted advertisements is huge (more than 100 million views per day in YouTube and reportedly over a billion for MySpace).[5] Also, information that reveals users' social networks, together with content-based networks and combinations of both, may be considered valuable.

[5] From an *Information Week* article: "MySpace Dethrones Yahoo As Page-View Leader", http://www.informationweek.com/news/showArticle.jhtml?articleID=196603894&subSection=

Metadata may be playing a crucial role in targeted advertising. It enables marketing on the basis of metadata entries, much as Google is currently doing with their AdWords, but using content metadata instead of web pages as the source. For instance, listening to Schbert's latest song on your MP3 player suggests that you might be interested in the ads by a clothing store he endorses.

As discussed in section 7.4., some online communities are more than virtual. This is the case with the online game *Second Life*, where virtual assets are purchased with real money.[6] It is worth noticing that even some (real) artists have started streaming "live" concerts in this virtual world.[7]

At the same time, virtual characters for the popular multiplayer game *World of Warcraft* are being offered in online auctions for bids in real money. One eBay seller listed *"Three level 60 characters (Warlock, Priest, Hunter)"* for bidding, justifying the $190 price with *"This levelling up three characters took me like more than 80–90 days. I want to sell this off for a cheap and reasonable price."* The potential of virtual economy may far surpass our expectations.

8.4   Obstacles in Reaching the Vision

Making a vision become reality requires a lot of research and development. What it also includes is overcoming a number of problems, threats, and obstacles along the way, be they technical, psychological, economical, physical, social, etc. In this section, we analyse some of the most probable issues along the way towards the vision. We divide the treatment into two categories: technology-originated issues, and human-related.

8.4.1   Technical Problems and Challenges

Most of the identified problems can be, not surprisingly, derived from the vision topics and technical enablers discussed above. For instance, the community-created ontologies, or any other ontologies, provide a good foundation for more generic ontologies. However, how far should we go? Should the whole world be modelled, and to what extent and granularity level? Such a task is close to impossible, not to speak of

---

[6] "The average going rate for land parcels in *Second Life* is between $100 and $1,000." Quoted from: "Second Life realtor makes $1 million" in http://www.gamespot.com/news/6162315.html

[7] http://www.wired.com/news/technology/internet/0,71593-0.html?tw=rss.index

maintaining the information. A more realistic approach is to build partial ontologies in the most important domains, and start to build the semantics based on these.

Another crucial issue is metadata interoperability. In case the existing numerous metadata formats continue to be, more or less, incompatible in terms of both definitions and semantics, the interoperability between applications and cross-media content remains a goal out of reach. An efficient mechanism for conversions between formats, as well as standardized ontologies, is an essential prerequisite for metadata to really start gluing the content objects together.

The quality of metadata is another significant factor that is reflected both in technological and user-related categories. Fragmented metadata is a particularly challenging issue. When some metadata exists, but is scattered sparsely across personal content, it provides too little information for solid automation. In a similar manner, cross-media, cross-application, and cross-manufacturer metadata formats and standards (that is, universal metadata management) are a challenge, since dozens of players from various domains are involved.

Context recognition and processing technologies may develop more slowly than anticipated, which will lead to starvation of context-hungry applications. This is a significant threat, since without file context (section 4.7) many features of smart metadata, relations, and social aspects become harder to implement.

Problems may also arise from unexpected directions. For instance, the increasing growth rates of personal content, and the expected growth of related metadata, require an increasing amount of mass storage and memory. We are used to thinking that the price of storage (bits per dollar) will keep decreasing but, for instance, a single fire in a chip foundry can significantly disturb chip production, drive flash memory prices up, and thus slow device sales to a (temporary) halt.

8.4.2 Human-Related Issues

One of the greatest challenges related to users and metadata is the fear of compromised privacy. Once the amount of personal content, together with potentially extremely intimate metadata, increases and spreads across devices, it will probably raise fears and uncertainty: how well will the information stay private? This is a justified concern, and should be addressed both in system design (that is providing sufficient methods for filtering metadata, encrypting it, and so forth) and user interface design (that is, making it explicit that the metadata and personal information is indeed kept safe and sound).

One typical human fear is related to automation: what if automatic synchronizing, archiving, and backing up should fail? What if it loses something irreplaceable, unique, and personally significant? This fear is well justified, given the unreliable and often unfathomable nature of computing devices. Could I trust my invaluable content to such a fragile system, one might ask. The land of permanently lost bits is alarmingly close to us.

Another human threat is the users who intentionally abuse metadata. There are several potential misuses, such as spying individuals from anonymous metadata by, for instance, combining information from a number of sources. Another potential threat is metadata forgery and spam – the users will create and distribute bogus metadata, thus potentially ruining the work of others, rendering a lot of metadata useless, and also interfering with the use of clean metadata (this can be compared to the current spam problem in e-mail).

A typical example, familiar from the past, is the misuse of adult web site metadata tags to include arbitrary keywords that bear no relation to the offered services. This has led the search engines, and subsequently the web surfers, astray. Not surprisingly, the metadata tags are not used by current search engines. Such forgery will undoubtedly plague metadata related to personal content as well.

As a side effect of metadata forgery, false metadata could lead to malfunctions through incorrect assumptions on people's needs, wishes, and future actions. False metadata can also potentially affect content usage. Once more content is protected in one way or another (see section 3.5 for discussion on Digital Rights Management), false rights information may render content object useless, or even illegal in the worst case scenario.

An interesting potential indirect threat is assuming that people's ability to make associations decreases due to reliance on automatically generated associations. This is a familiar line of argument that inevitably comes up with every new technology. Initially it is feared that the technology will take over our humanity and reduce our brain capacity. Such discussions have been raised with TV watching and the ability to read, for instance. Too much automation and machine operations may indeed weaken some of our mental abilities. Alternatively, it could be argued that the human mind will be released for more interesting tasks, while machines do the low-level associating.

Then there is always the issue of public acceptance. In case the use of metadata remains too complicated or is not regarded useful, people will (rightfully so) find other solutions for their personal content needs. However, the reasons for such non-acceptance should be identified and analyzed. For instance, it may also be due to fact that the reasoning related to automated operations is not visible to users – they cannot understand the relation between the starting point and the result. We may need explanation mechanisms for metadata.

One threat that is closely related to automation, user interfaces, and human acceptance, is the proper level of metadata visibility. In other words, how to make it predictable what some metadata-enhanced action will produce? How to make it perceivable to the user what metadata attributes, values, automation, inference, and so forth has resulted in the current view or state.

Even though we have discussed how metadata aids managing content, there is another perspective that would merit further study: how could metadata help people to feel better? How could we employ metadata in a manner that makes systems dealing with personal content more *calm* (Weiser and Brown 1996)? These issues are not yet well understood and are being studied in various fields, for instance from the viewpoint of *affective computing* (Picard 1997). We believe that personal content will have a higher emotional value to a user, and therefore may serve better in affective systems than unfamiliar content.

Technology will progress and develop at a pace that easily surpasses many human capabilities. Therefore, we need to seek out new user interface and interaction paradigms and technologies that aid in personal content experiencing while on move. For instance, people cannot pay a lot of visual attention to a device when walking on a busy street, which means that the mobile device UI should support rapid *glance and click* type of interactions that do not require more than couple of seconds to convey all relevant information. This, of course, requires a lot of technology in the background to select what is relevant in the current context.

8.5   From Databases to Lifebases

Many visions of personal content to come include the notion of a *memory prosthesis*, that is, a system that records and manages everything; one example is MyLifeBits (Gemmell et al. 2006). This typical memory prosthesis includes a camera attached to your head-worn display, which is always on, always recording: all you see is seen and stored by the camera – for as long as you are wearing such headgear (Figure 8-2). Likewise, your auditive environment is continuously recorded. A multitude of other data may also be recorded, such as location, physical parameters, sensor readings, and so forth. A mobile phone can also be geared towards a memory prosthesis (Rhee et al. 2006).

Figure 8-2. A head-worn display attached to a wearable computer prototype. A camera for the memory prosthesis can be installed in the display frame (Photo: courtesy of Tero Hakala).

This way, once sufficiently efficient search mechanisms are in place, you are able to access any piece of your personal history (not just single fragments of it). Not that all recorded information is meant for accessing, as a lot of our daily routines may bypass without us wishing to re-live any part of it, but still the content can be used to deduct patterns in daily behaviour. This helps in anticipating your future actions and adapting your computing environment accordingly.

The various memory prosthesis visions and prototypes differ in their recording devices and retrieval principles. Some record video, some locations, yet others rely on capturing our electronic environment. Some are carried at all times, and some reside in desktop computers. Some record everything, while some only store events that the user explicitly indicates.[8]

What is common to all these visions is our notion of *timeshifting life* – storing experiences for later use. This is the ultimate form of personal content, about as encompassing as it gets. Think of your *lifebase* as the key place for accessing all your content, organized into experiences.

---

[8] A good collection of links on the topic is available in: http://www.sigmm.org/Members/jgemmell/CARPE

```
COMMENT: AN EXCERPT FROM THE HOMEPAGE OF LARRY RUDOLPH

I am currently thinking about cell phones and controlling
my own personal information. This includes searching,
security, privacy, prediction, managing multiple devices,
and interacting with distributed data, services, and
people.

I mostly remember stuff based on time, place, and event.
I program my phone (in python) to tag all content I gen-
erate based on these attributes. I also try to get these
attributes into human terms. So, time is "working",
"relaxing", and so on. Location may be "home", "work",
"library", "Boston", "shopping", and so on. These are my
labels but my system actually learns them from my calendar
and blog entries (my daily activities are automatically
included into my blog or diary). I can find things using
a query such as: "who called me when I was listening to
the Beatles on my way home from work last week?"

Now that I have a model of my daily travels, I can lie
about my location. It is mostly correct, but no guaran-
tees. Since I control the distribution of my personal
information, I can choose what to reveal and what to hide.
This is a very powerful way of controlling one's personal
information.

Professor Larry Rudolph
Principal Research Scientist
MIT
```

(Professor Larry Rudolph's current research and teaching focuses on technologies and applications of experimental pervasive computing. http://csg.csail.mit.edu/u/r/ rudolph/public_html/)

We are experimenting with various approaches to life retrieval. One of our research projects, Sharme,[9] emphasises sharing of memories (the fundamental bits of life) with other people. It records life events and data with an embedded mobile phone application, which then transfers the data to a server (e.g., the user's PC) for subsequent processing, storage, and retrieval.

Figure 8-3 shows a prototype of the Sharme retrieval system. In the figure, the system has generated rudimentary summaries of someone's photos and locations during a visit in Italy. The lifebase is shared with

---

[9] http://www.sharme.org/

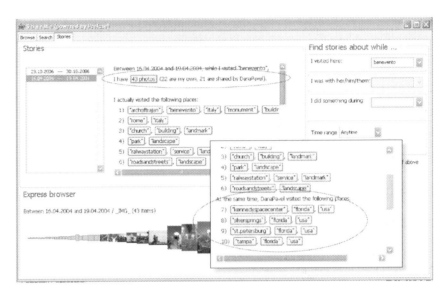

Figure 8-3. A PC prototype of the Sharme memory retrieval system (reproduced by permission of © Nokia).

another person, who had at the same time been on a holiday trip in Florida. The internal representation is a semantic network with an ontology about life events, in this case, holidays. We plan to add a presentation generator to the prototype, which allows us to create rich summaries of experiences, perhaps in the form of multimedia presentations, or automatically created wish-you-were-here postcards.

The Sharme system gives another dimension to sharing of personal content. So far, personal memories have been shared via more traditional means: postcards, letters, pictures, videos, blogs. Automated means for collecting the memories have not been in place, not to speak of organizing and sharing that data. We are experimenting with various options of organizing personal life data for efficient recall, synchronizing, backups, and other GEMS actions. Memory prosthesis makes it possible to treat our life data as just another kind of personal content.

There are obvious concerns with such life sharing. Who will have access to my lifebase? What am I willing to share? What will I not share? What will I really like to remember, and what should rather be forgotten? How to find the important bits? Is it really worth trying to keep everything? How could we summarize life events? And so on. All these questions are being debated on various fronts, and the personal content boom makes these questions ever more timely.

While we are not ready to answer all these questions yet, one thing is certain: memory prostheses are already there, in your digital cameras, phones, PCs, and any number of electronic devices that timeshift your life data. The information is not yet collected, organized, and GEMSed as a whole, but it will be. With the help of metadata.

ERASE AND REWIND

A week passes. Eddie and Angie argue endlessly. Another week, another fight.

"But you're selling yourself!"

"Thats none of your business!"

"It bloody is! You're my girl!"

"Think I am? Think again."

Angie grabs her gadget-laden purse, steps out and slams the door.

Eddie just sits there for the longest time. The sun sets, throwing irregular shadows on the wall.

Women.

How did it ever get to this? They were so close.

Eddie just sits there, staring at the window, seeing nothing. The lamp pops a thousand blobs.

Then, after quite some time, he goes to his computer, and starts the process of undoing their past. Search everything Angie in his lifebase. Erase it. Erase and rewind. Start over.

It's going to be a long process.

8.6   Move that Metadata!

(Excerpt from the keynote speech by Harry Hacker, President of the Context Jockey Society, in PCE '2012 – International conference on Personal Content Enjoyment, Phoenix, Arizona, October 2012)

"All right. It's time for you guys to get moving. Start collecting that metadata, now. It's not rocket science, just a bunch of descriptors on everyday data. Just make sure you try to include all the relevant contextual and social data you can, if you dare, given the privacy risks. If it's just your own lifebase you are populating, don't worry. And, by the way, never delete anything. The falling mass storage prices will let you just keep on adding data for years and years.

"Meanwhile, stand by while we are developing the machinery for the metadata magic. The Web is ready, search engines are ready, mobile computing is ready. All we are waiting for is a few agreements on metadata standards. Plus a few pieces of software to process that metadata and craft out new metadata from existing content. The prototypes are there, but it will take us a couple years to hone those systems such that we can start to offer Internet services with them. Then it will be a few more years before that will happen in your pocket, in your mobile devices. But it will. Because it can.

"If you are sitting on a pile of interesting content, go ahead, share it. If it happens to be yours, something you made, no problem. Just grab a Creative Commons licence and put it out there for others to enjoy. If you do not actually own the content, share it anyway – but in the form of links, ratings, recommendations, not the actual content. While you are at it, attach a micro-donation scheme to your metadata and wait for those e-coins to fill your bank account. Or, then not. Maybe you'll be happy with that good old currency: fame. You will get that, because there's room for everyone in the Long Tail.

"Then if you are interested in a slow but sure income, get a jockey job! Go out there into the world, get among the people, record your context, and upload it into those data mining centres. They can use everything you upload there. Better yet, start annotating the world around you with semantics. Or if you are not a computer scientist, it's okay just to tag it. The wisdom of the crowds will sort it out.

"Now, if you all would be so kind and turn on your experience engines for a moment . . . thank you . . . I want to make a statement and be sure it gets recorded. So, here's the deal. If the complete Semantic Web will not be in your pocket in two years – that's by the end of 2014 – I will eat my hat. Promise.

"So, that's what I have to say. And one more thing to all those jockeys running on the streets – let's be careful out there!"

8.7   References

Gemmell J., Bell G. and Lueder R (2006) MyLifeBits: A Personal Database for Everything. *ACM Communications*, Volume. 49, Number 1, Jan., pp. 88–95. ACM Press, New York, USA.

Picard R.W. (1997) *Affective Computing*. MIT Press, Cambridge, MA, USA. 275 p.

Rhee Y., Kim J. and Chung A (2006) Your Phone Automatically Caches Your Life. *ACM interactions*, Volume XII, Number 4, Jul./Aug., pp. 42–44.

Weiser M. and Brown J.S. (1996) "Designing Calm Technology", *PowerGrid Journal*, v 1.01, http://powergrid.electriciti.com/1.01

Epilogue

Personal Content Experience: Managing Digital Life in the Mobile Age J. Lehikoinen, A. Aaltonen, P. Huuskonen and I. Salminen © 2007 John Wiley & Sons, Ltd

SECOND LIFE

Bob thought "Circus" was a weird name for this roundabout, which was not even a roundabout, just a complex intersection of five streets. He stood there observing the strange combination of neon signs, historic buildings, and just impossible traffic. Which was not his main topic today, of course. He was studying the area to prepare for the – you guessed it – upcoming demonstration against the multinationals. He did not like the idea of protesting in London, their AARGH.org cell's homebase, but you do what you have to do. Those globalists must be stopped.

There were five tube exits. Blocking those with smell bombs would be nasty, but it would just cause a major disturbance without really hurting people. Blocking the traffic by blowing a couple of double-decker bus tires might make some people quite angry, but again it was better than driving the bomb trucks in. There would be lots of attention with their slogans flowing around in real and virtual spaces. In addition, they planned to take over the video displays and plant new content there, and the same with local Wi-Fi, 3G, and DVB towers. It would be a major attack, and likely to get them to jail, but could be worth it.

Bob saw tourists photographing the square. Funny – a man in coveralls was shooting a photo of the Eros statue. A lot of those pictures would end up on the Net, which was good for them. Martin, the 3D modeller, was already running the stitcher software that takes those pictures and videos, correlates them in space, and creates a viewable model that could be viewed in 3D from arbitrary angles. They already had major parts of the square modelled. Now Bob would have to walk through the subway system to get video footage of the tube exits, so they could complete the model in the comfort of their own computers. Armchair activism . . .

Bob turned around and saw yet another 3Lifer there in the corner. A bunch of those there today. What a weird lifestyle. All those spacer people with their silly phones in their awkwardly extended hands, peering at the world through a looking glass. And they had no idea. Their virtual worlds did not give them a clue as to what was really going on in the real world. Which was broken by the greed of those multinational supercorporations.

Time do something about it.

FIRST LIFE

Cathy and Deena had agreed to meet at the 'Circus. Deena had some business meeting in Coventry Street, so it would be a short walk for her, and Cathy just liked walking along Haymarket and then stopping at the Memorial fountain. She sat there by the fountain, a little away from the men at work who were fixing something on the other side.

There were lots of people, as usual. Cathy loved watching people. Early in her career she had tried doing portraits, but that was not her thing. The herbs were more beautiful, in a weird way more personal, more alive than humans. She thought she would not be able to exhaust that area in a hundred years, even though the archives showed that she had already scanned more than 2 000 of her works on that topic, with more waiting in the attic.

She did like people in general, though a little less intensely those folks with their virtual glasses, or even those peeking at the world through their phones. Like that kid there in the corner, chuckling to himself as he pointed his phone around. What was it that drove people into 3space? Was the real world already not good enough? Just look at that marvellous sophisticated rich tapestry of human life everywhere! Why would anyone want to escape to that clean, sterile, cold virtual life, was simply beyond her.

Deena's work was different. Even though she was building 3D models, she made a point of not including humans, not a single silly blocky pixely avatar, and somehow this very thing made her buildings appear all the more humane. Strange, but that's how it worked.

Still no Deena. Not to hurry. More 3lifers there, crossing the street. Steve had proposed that Deena and Cathy get busy designing those virtual spaces for the 3lifers, which could even mean quite some income. Deena seemed half curious but Cathy had dismissed the idea. She did not want to touch that piece of crap. Their web shop was already enough to worry about. And . . . oh no – the backup! She had forgotten to ask Steve about the backup. Her PC had reported an error upon her last scan job. What should she do? Steve, help!

GOOD OLD LIFE

Dave was not in his best mood. Fixing pipes never was a clean job, but this was getting bad. This fountain in Piccadilly Circus was simply filthy. Nice Eros statue on top, but the bottom of the fountain had not apparently been properly cleaned in years, and the standing water, bird manure, and leftover hamburgers from tourists had developed into a hideous slime that would probably kill anything that touched it. And he had to fix the piping there, with all those tourists watching and making fun at him.

He should really work on his new betting site to be able to retire soon. Need to talk to Eddie, but he was a busy kid these days.

Those pipes were really old, probably from the 19th century. How old was the memorial anyway? 1880? Dave remembered reading about the Eros statue in an article about Piccadilly Circus. His granddaughter had prepared a history essay for her school, and he had helped her to find material. Much of the history was still on paper, which felt natural for Dave. His granddaughter was more at home on the Internet. Dave showed her how it was possible to actually go and browse a paper-based index (the kid had waved her little head in disbelief) and retrieve, old, worn, physical paper copies of paper-based magazines in a library.

Dave replaced the regulator valve and secured the joints. Now there. That should do it. Tomorrow they could actually test for more leaks. Could call it a day now. He got his gear, stepped out of the fountain, waved goodbye to the tourists and got a small applause that was practically drowned by the noise from traffic and tourists.

The winged Eros statue on the top of the fountain was truly beautiful. At a moments thought, Dave got his cellphone and took a picture of the statue. Would be nice to show Esther how he was doing a worthy deed, restoring the landmark to its former glory.

Index